大肠杆菌
感染致奶牛乳腺炎的致病机制及防治

庄翠翠 / 著

中国纺织出版社有限公司

内 容 提 要

大肠杆菌是奶牛乳腺炎重要的致病菌。目前，抗生素是国内外治疗奶牛乳腺炎最常用的方法，然而大量不合理使用抗生素，使耐药致病菌株不断增加，威胁人类公共卫生安全。本书阐述了奶牛乳腺炎的定义及分类、奶牛乳腺炎的病因、牛奶中的体细胞数、乳腺炎模型的建立、*E. coli* 感染致奶牛乳腺炎的致病机制，以及大肠杆菌型奶牛乳腺炎的防治。本书可为预防和控制大肠杆菌型奶牛乳腺炎的发生提供理论依据，以期保护人和动物的健康、保证食品质量、保障公共卫生安全。本书适合高校教师、学生以及相关专业研究人员阅读。

图书在版编目（CIP）数据

大肠杆菌感染致奶牛乳腺炎的致病机制及防治 / 庄翠翠著. -- 北京：中国纺织出版社有限公司, 2025.3.
ISBN 978-7-5229-2620-9

Ⅰ. S858.23

中国国家版本馆 CIP 数据核字第 2025YR2057 号

责任编辑：范红梅　　责任校对：王花妮　　责任印制：王艳丽

中国纺织出版社有限公司出版发行
地址：北京市朝阳区百子湾东里 A407 号楼　邮政编码：100124
销售电话：010—67004422　　传真：010—87155801
http://www.c-textilep.com
中国纺织出版社天猫旗舰店
官方微博 http://weibo.com/2119887771
三河市宏盛印务有限公司印刷　各地新华书店经销
2025 年 3 月第 1 版第 1 次印刷
开本：710×1000　1/16　印张：12.25
字数：192 千字　定价：78.00 元

凡购本书，如有缺页、倒页、脱页，由本社图书营销中心调换

前　言

奶牛乳腺炎对奶牛养殖行业危害较大,严重影响奶牛的健康,导致产奶量大幅下降,直接影响奶农的经济收益。患病奶牛的乳汁质量受损,可能含有病原菌等有害物质,对乳制品安全构成威胁,进而影响消费者健康。此外,乳腺炎的治疗成本较高,增加了养殖成本,还会使奶牛淘汰率上升,造成资源浪费。同时,它对奶牛养殖业的可持续发展也带来挑战,可能引发一系列产业问题,如市场波动、行业发展受阻等,严重制约奶牛养殖业的健康稳定发展。据估计,我国每年因奶牛乳腺炎造成的损失约高达1.35亿元人民币。乳腺炎不仅影响产奶量,造成严重的经济损失,而且影响乳的品质,危害人类健康,其防治问题已日益为人们所关注。

大肠杆菌是我国流行率较高的环境型致病菌之一,能够在分娩前、后和泌乳早期引起奶牛乳腺炎。在英国,大肠杆菌感染是引起奶牛乳腺炎最常见的原因;在荷兰,大肠杆菌引起的临床型乳腺炎发生率较高;在美国,大肠杆菌感染引起的乳腺炎是奶牛养殖业高发的疾病之一;在中国,大肠杆菌是奶牛养殖场最常见的病原菌,占14.4%。大肠杆菌持续感染引起的奶牛临床型乳腺炎表现为从轻度的局部症状到严重的临床症状并累及全身,甚至导致奶牛的死亡,但其受到的关注较少。因此,研究大肠杆菌型奶牛乳腺炎的致病及防治机制对奶牛养殖业有重大意义。

在治疗大肠杆菌型乳腺炎时应选取快速且有效的治疗策略,如抗生素的使用。当奶牛发生急性大肠杆菌型乳腺炎时,奶牛防御机制的失败会导致乳腺内细菌的无限增长,此时抗生素治疗有助于控制感染,且增加患牛的存活率。目前,β-内酰氨类药物是国内外兽医治疗大肠杆菌型临床乳腺炎的首选策略,然而β-内酰氨类药物的大量使用诱使大肠杆菌产生超广谱β-内酰胺酶。超广谱β-内酰胺酶通过在普通的β-内酰胺键中加入一个水分子以水解氧亚氨基头孢菌素,包括头孢噻肟、头孢曲松和头孢他啶等第三代头孢菌素的β-内酰胺酶。因此,在我国患有乳腺炎的奶牛中产β-内酰胺酶大肠杆菌的高检出率可能与持续使用抗生素和(或)较少控制使用抗生素治疗乳腺炎有关,这进一步突显了进行大肠杆菌型奶牛乳腺炎非抗生素替代疗法研究的紧迫性和重要性。

基于此,笔者将多年的科研工作积累成书,旨在增强人们对乳腺炎的了解。作者以自己多年来从事奶牛乳腺炎研究的成果为基础,通过查阅相关的文献资料,参考了国内外同行专家和学者的部分研究成果,系统阐述了大肠杆菌感染致奶牛乳腺炎的致病机制及防治方法,为预防和控制乳腺炎、保护公共卫生安全提供参考。

全书共分三章,第一章介绍奶牛乳腺炎概况;第二章介绍大肠杆菌感染致奶牛乳腺炎的致病机制;第三章介绍大肠杆菌型奶牛乳腺炎的防治机制。本书采用组织学、生物化学、分子生物学、细胞生物学及分子毒理学等手段,通过动物在体实验和体外细胞培养实验,深入研究了大肠杆菌感染对奶牛乳腺炎的的影响。本书每个章节后都附有主要参考文献,以供读者查阅和追踪。

由于作者水平有限,书中疏漏之处在所难免,恳请同行专家以及广大读者批评指正。

庄翠翠
2024 年 7 月
于山西农业大学

目 录

第一章 奶牛乳腺炎概况 …………………………………………… 1
    第一节 奶牛乳腺炎的定义及分类 ………………………………… 1
    第二节 奶牛乳腺炎的病因 ………………………………………… 9
    第三节 牛奶中的体细胞数 ………………………………………… 75
第二章 大肠杆菌感染致奶牛乳腺炎的致病机制 ………………… 94
    第一节 大肠杆菌感染诱导免疫应答的机制研究 ………………… 94
    第二节 大肠杆菌感染诱导乳腺上皮细胞焦亡的机制研究 …… 110
    第三节 大肠杆菌感染诱导线粒体损伤的机制研究 …………… 126
    第四节 大肠杆菌感染诱导乳腺上皮细胞
           铁死亡的机制研究 ……………………………………… 136
第三章 大肠杆菌型奶牛乳腺炎的防治机制 …………………… 152
    第一节 番茄红素缓解奶牛乳腺炎的机制研究 ………………… 152
    第二节 硒缓解大肠杆菌感染致炎症的机制研究 ……………… 166
    第三节 硒缓解大肠杆菌感染致细胞凋亡的机制研究 ………… 178
后 记 ……………………………………………………………… 188

第一章 奶牛乳腺炎概况

第一节 奶牛乳腺炎的定义及分类

一、奶业的发展现状

奶业作为现代农业和食品工业的重要组成部分,对于居民膳食结构的优化、国民体质的提升,以及农牧民收入的增长都具有深远意义。据联合国粮食及农业组织(FAO)统计,印度、美国、巴基斯坦和中国这些奶产量大国,其产奶量均呈现上升态势(表1-1)。自2008年以来,我国各地、各部门严格贯彻党中央国务院的部署,将保障乳品质量安全作为核心工作,全方位开展乳品质量安全的监督执法和专项整治行动,加快转变奶牛养殖的生产方式,推动乳品加工的优化升级,奶业素质显著提高,现代奶业建设成绩显著。奶业已然成为现代农业和食品工业中最具活力、发展最快的产业之一。奶业全产业链的质量安全监管体系日益完善,监管力度不断加大。生鲜乳抽检覆盖了所有奶站和运输车,乳制品实行出厂批批检验制度。从2008年至今,累计抽检生鲜乳15.1万批次,整顿奶站11893个,奶站的基础设施、卫生状况和检测条件等均有明显改善。2015年,生鲜乳中的乳蛋白、乳脂肪抽检平均值分别为3.14 g/100 g、3.69 g/100 g,均高于《生乳》国家标准,规模牧场的相关指标达到发达国家水平;违禁添加物抽检合格率连续7年保持100%。乳制品抽检合格率达99.5%,婴幼儿配方乳粉抽检合格率为97.2%。大力推动奶牛标准化规模养殖,实施振兴奶业苜蓿发展行动,开展奶牛遗传改良计划,奶牛养殖的规模化、标准化、机械化和组织化水平大幅提升。

表 1-1　2020 年全球奶产量情况

项目	印度	美国	巴基斯坦	中国
奶产量（亿吨）	1.95	1.01	0.486	0.354
同比增长（%）	2.1	2.1	2.7	7.4
全球奶产量占比（%）	22.7	11.8	5.6	4.1

2015 年，我国 100 头以上奶牛的规模养殖占比达 48.3%，相较 2008 年提高了 28.8 个百分点。机械化挤奶率达到 95%，提升了 44 个百分点，规模牧场已全部实现机械化挤奶。泌乳奶牛年均单产达 6 吨，增加了 1.2 吨。规模场全混合日粮饲养技术（TMR）的普及率达 70%。奶农专业合作组织超过 1.5 万个，是 2008 年的 7 倍有余。2015 年，规模以上乳制品企业（年销售额 2000 万元以上）有 638 家，比 2008 年减少 177 家，婴幼儿配方乳粉企业 104 家，比 2011 年减少 41 家。奶业 20 强（D20）企业的产量和销售额占全国 50% 以上，有 2 家企业跻身世界乳业 20 强之列。自 2008 年起，国务院及相关部门相继颁布施行《乳品质量安全监督管理条例》《奶业整顿和振兴规划纲要》《关于进一步加强婴幼儿配方乳粉质量安全工作的意见》《乳制品工业产业政策》《推动婴幼儿配方乳粉企业兼并重组工作方案》和《婴幼儿配方乳粉产品配方注册管理办法》等 20 多项规章制度，公布《生乳》国家标准等 66 项乳品质量安全标准，推出促进奶牛标准化规模养殖、振兴奶业首蓿发展行动、奶牛政策性保险、乳品企业技术改造、婴幼儿配方乳粉质量安全追溯等重大政策，初步构建起覆盖全产业链的政策法规体系。中国奶业发展迅速，态势良好，呈现出罕见的积极趋向。2020 年，中国奶产量达 0.354 亿吨，跃居全球第四，约占全球奶产量的 4%，其中牛奶产量约为 3440.0 万吨，同比增长约 7.5%；2021 年，我国奶业发展保持良好态势，在提质增效方面开创了新局面，2021 年牛奶产量 3683 万吨，同比增长 7.1%（图 1-1）；2022 年，中国原料奶产量增加，乳制品进口大幅降低，奶源自给率实现近 7 年来的首次回升，中国牛奶产量达 3932 万吨，同比增长 6.8%（图 1-2），创下历史新高。此外，2020 年我国实现了自 2014 年以来奶牛存栏量持续减少后的首次增长，奶牛存栏量达 506.0 万头，截至 2021 年年末，全国荷斯坦奶牛存栏 561.2 万头，同比增长 10.9%（图 1-3）。同时，2020 年我国奶牛年单产量约 8.3 吨，2021 年全国荷斯坦奶牛平均

年单产为8.7吨,同比增长4.8%(图1-4),刷新了我国奶牛单产的历史纪录。处于"两个一百年"的交汇节点,中国奶业不断发展壮大。同时,奶牛养殖业的机械化、信息化、智能化水平显著提升,进一步推动奶业健康、持续地发展。

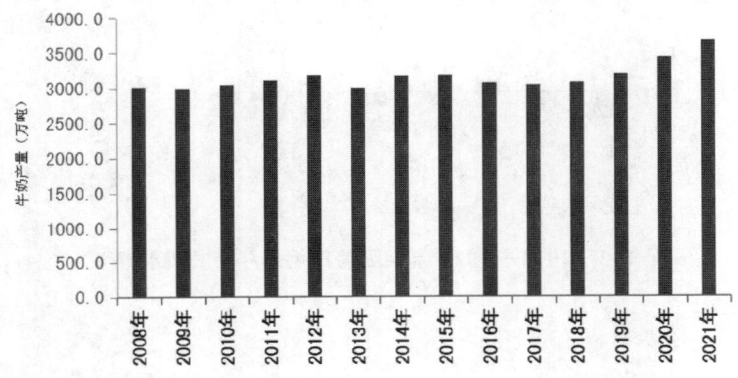

图1-1 2008—2021年中国牛奶产量变化情况

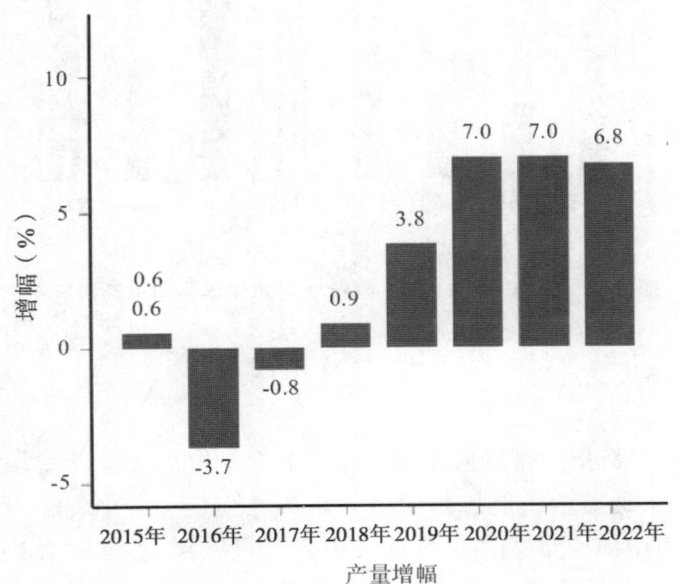

图1-2 2015—2022年中国牛奶产量增幅

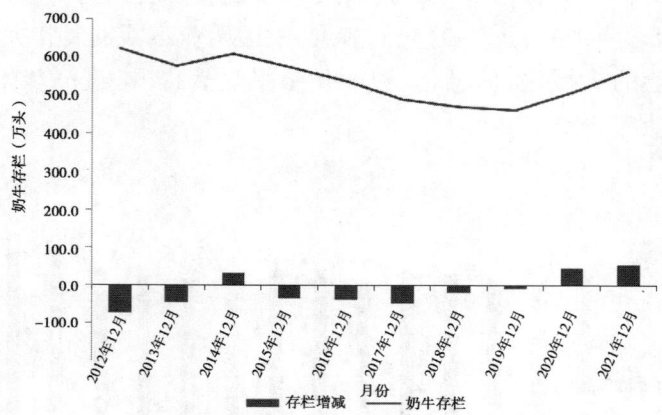

图1-3 2012—2021年全国荷斯坦奶牛存栏变化情况

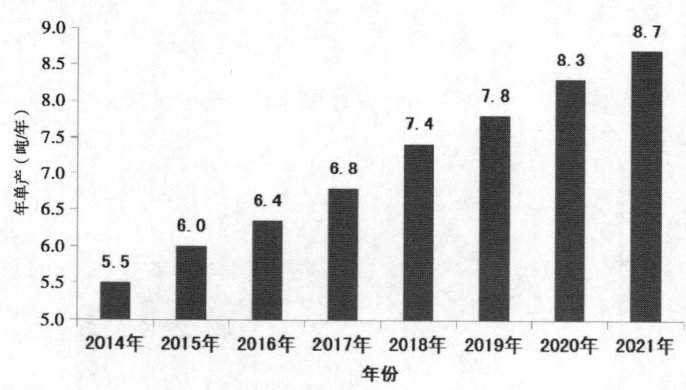

图1-4 2014—2021年全国荷斯坦奶牛年平均单产情况

二、乳腺炎的定义

奶牛乳腺炎指的是乳腺组织在受到物理、化学、微生物等刺激后所产生的炎症，通常会给乳腺组织带来永久性损伤。奶牛乳腺炎在奶牛规模化养殖中属于最为常见且危害极为严重的疾病之一，会致使奶产量下降、治疗费用增多、饲料利用出现损失、奶废量上升、奶牛成熟前淘汰率提高，不但对产奶量产生严重影响，还会极大危害奶品质，造成巨大的经济损失，威胁人类健康。奶牛乳腺炎在世界各国的奶业养殖中都有着较高的发病概率,是制约奶业健康发展、提高奶业质量的最大阻碍。有报道称,奶牛乳腺炎是乳品行业里最为常见、总体花费最为高昂的疾病,并且具有全球性关联。据相关统计,全球约30%的奶牛患有各类乳腺炎,每头奶牛因奶牛乳腺炎所导致的经济损失约为147美元,全球每年的损失约达350亿美元,其中在美国达20亿美元,乳腺炎每年给新西兰乳业造成的经济损失超过3亿新西兰元。全球奶牛乳腺炎的发病率在30%~50%,其中美国的奶牛乳腺炎发病率约为50%,俄罗斯约为38%,西欧部分国家奶牛隐性乳腺炎的患病率为25%,日本奶牛乳腺炎的平均发病率为45.1%。相较国外而言,我国奶牛乳腺炎的发病状况更为严峻,在20%~75%,部分严重地区甚至超过75%。据统计,奶牛临床型乳腺炎的阳性率平均为33.41%,隐性乳腺炎患病率平均为46.40%~85.70%。在我国,每年因奶牛乳腺炎这一问题造成的经济损失高达30亿元人民币。

三、奶牛乳腺炎的分类

奶牛乳腺炎一般分为两大类:临床型乳腺炎和亚临床型乳腺炎。临床型乳腺炎可通过牛奶、奶牛乳房和奶牛的其他变化发现;亚临床型乳腺炎则变化微小,需要借助实验室检验进行检测。二者之中,亚临床型乳腺炎症状隐蔽,难于诊断,防控难度大,发病率和造成的经济损失均大于临床型乳腺炎。

(一)临床型乳腺炎

发生临床型乳腺炎的奶牛乳房往往会有疼痛、发热、肿胀的症状,甚至会变得坚硬以及受损,泌乳减少乃至停止,乳汁的性状会出现可视性改变,比如变色、出现凝块等,严重时会出现血样乳,还可能伴随明显的全身症状,如情绪抑郁、拒绝进食甚至死亡。依照病程的长短和病情的严重程度,临床型乳腺炎又能分为最急性乳腺炎、急性乳腺炎、亚急性乳腺炎、慢性乳腺炎。

在奶牛乳腺炎急性发作时期,触摸乳房能感觉到温度上升,外观肿胀且常伴有非正常的水样分泌物。有些病例还会同时引发严重的系统性疾病,常在外观症状出现之前或者牛奶样品出现可视性变化之前产生,败血症或毒血症是其主要的致病原因,症状涵盖发热、瘤胃蠕动缓慢、食欲不佳、脉搏加快、腹泻、情绪抑郁等,当身体状况进一步恶化时可能致使病牛脱水死亡。在乳房出现中度至重度的炎症发作时,急性奶牛乳腺炎最为多见。此时牛奶常常呈现凝块状的水样,同时产奶量显著下降。其系统性症状与最急性奶牛乳腺炎相似,但相对严重程度稍小,症状较为缓和。亚急性奶牛乳腺炎常见于轻微的炎症,乳房的外观通常没有可视性的变化,可在患牛的牛奶样品中发现小的凝块或絮状物及轻微的颜色改变,不会产生系统性疾病和症状。慢性乳腺炎是指乳腺炎反复出现的情况,可能表现为亚临床型乳腺炎,并能够在数月数年的时间里多次发作,长此以往会造成乳腺组织的永久性损伤。乳房会出现脓疮、硬化、瘢痕等具体变化,触感变硬,左右不对称,严重影响奶牛的生产性能,最终只能淘汰患病奶牛。

(二)亚临床型乳腺炎

亚临床型奶牛乳腺炎相较于临床型奶牛乳腺炎更为多见,其发生率比临床型高 15%~40%,不会呈现显著的乳腺炎症状,牛奶产量下降、质量变差,且很少有乳汁的外观变化,有时需要借助特殊的实验手段来加以检测。亚临床型奶牛乳腺炎的乳汁中的体细胞、pH 值和导电率等通常偏高。倘若畜主未能及时察觉,可能会延误治疗,亚临床型奶牛乳腺炎或许会转变为临床型乳腺炎,进而对奶牛的身体健康和生产性能产生影响。美国乳腺炎协会(NMC)制订了专门的标准用于对其检测。其一的方法是通过检测奶样中的体细胞数量,当乳体细胞数为 20 万~50 万个 /mL 时,就能够诊断为亚临床型乳腺炎;其二的方法是加利福尼亚乳腺炎检测(CMT),将检测液与奶样混合后缓缓摇动,检测液里的十二烷基磺酸钠等表面活性剂能够使乳体细胞、乳腺炎致病菌等细胞中的 DNA 析出凝集,并以溴甲酚紫作为显色指示剂。依照颜色变化、混合液黏性、是否出现沉淀、沉淀是否易于分散等指标,将奶样划分为乳腺炎阴性、可疑、弱阳性、阳性、强阳性五个等级。

参考文献

[1] 刘长全.2021 年中国奶业经济形势回顾及 2022 年展望 [J]. 中国畜

牧杂志,2022,58（3）:232-238.

[2] 刘长全,张鸣鸣.2022年中国奶业经济形势回顾及2023年展望[J].中国畜牧杂志,2023,59（3）:307-315.

[3] 岳婷婷,谢书宇,高金芳,等.奶牛乳腺炎的防治研究进展[J].中国奶牛,2017（9）:38-42.

[4] SHARMA N, JEONG D K.Stem cell research: A novel boulevard towards improved bovine mastitis management[J].International Journal of Biological Sciences,2013,9（8）:818-829.

[5] HOGEVEEN H, STEENEVELD W, WOLF C A.Production diseases reduce the efficiency of dairy production: A review of the results, methods, and approaches regarding the economics of mastitis[J]. Annual Review of Resource Economics,2019,11:289-312.

[6] SATHIYABARATHI M, JEYAKUMAR S, MANIMARAN A, et al.Infrared thermography: A potential noninvasive tool to monitor udder health status in dairy cows[J].Veterinary World,2016,9（10）:1075-1081.

[7] 郝景锋.吉林省奶牛隐性乳房炎主要致病菌敏感中药筛选与初步应用[D].长春:吉林大学,2018.

[8] OLDE Riekerink R G M, BARKEMA H W, VEENSTRA S, et al.Prevalence of contagious mastitis pathogens in bulk tank milk in Prince Edward Island[J].The Canadian Veterinary Journal: La Revue Veterinaire Canadienne,2006,47（6）:567-572.

[9] HOGEVEEN H, HUIJPS K, LAM T J G M. Economic aspects of mastitis: New developments[J].New Zealand Veterinary Journal,2011,59（1）:16-23.

[10] 吴琼.乳酸杆菌减轻大肠杆菌诱发奶牛乳腺上皮细胞炎性损伤的机理[D].北京:中国农业大学,2018.

[11] HALASA T, HUIJPS K, ØSTERÅS O, et al.Economic effects of bovine mastitis and mastitis management: A review[J].The Veterinary Quarterly,2007,29（1）:18-31.

[12] 刘永夏.奶牛乳房炎大肠杆菌F17A-MF59疫苗免疫学特性研究[D].北京:中国农业大学,2015.

[13] ZHANG Z, LI X P, YANG F,et al. Influences of season, parity,

lactation, udder area, milk yield, and clinical symptoms on intramammary infection in dairy cows[J].Journal of Dairy Science,2016,99（8）: 6484-6493.

[14] WAGNER SA, JONES D E, Apley MD.Effect of endotoxic mastitis on epithelial cell numbers in the milk of dairy cows[J]. American Journal of Veterinary Research,2009,70（6）: 796-799.

[15] HULTGREN J, SVENSSON C.Lifetime risk and cost of clinical mastitis in dairy cows in relation to heifer rearing conditions in southwest Sweden[J].Journal of Dairy Science,2009,92（7）: 3274-3280.

[16] DRILLICH M, RAAB D, WITTKE M,et al. Treatment of chronic endometritis in dairy cows with an intrauterine application of enzymes[J].A field trial.Theriogenology,2005,63（7）: 1811-1823.

[17] 刘丕龙.奶牛乳腺炎来源的葡萄球菌甲氧西林抗性传播的机制研究 [D]. 杨凌：西北农林科技大学,2017.

第二节 奶牛乳腺炎的病因

奶牛乳腺炎的病因学研究不断增多,已从奶牛乳汁中分离出约150余种病原体。依据所涉及病原体的类别,乳腺炎大致能分为三类:细菌性乳腺炎、霉菌/真菌/藻类乳腺炎以及支原体乳腺炎。由于大多数乳腺病毒感染源自一般病毒血症,这是身体其他部位的病毒性疾病的典型特征,所以病毒作为致乳腺炎因子的临床意义极小。研究表明,能够致使奶牛乳腺病毒感染的主要有牛疱疹病毒、假牛痘(乳结节)病毒和牛痘病毒,其中牛痘病毒感染会引发溃疡性牛乳腺炎(乳腺脓疱性皮炎),真正受感染的是乳腺的表皮和真皮,而非乳腺的分泌泡。因此,如果不伴随细菌继发侵袭影响乳腺内组织,单纯的病毒性乳腺炎没有临床意义。综上所述,从病理学和经济学角度来看,细菌乳腺炎是最为重要的乳腺炎类型。

细菌型奶牛乳腺炎可被分为三种类型:传染型乳腺炎、环境型乳腺炎和机会型乳腺炎(图1-5)。其中,引发传染型乳腺炎的细菌(如金黄色葡萄球菌)能够在宿主身体内存活,特别是在乳腺组织内,进而引发亚临床症状,主要表现为感染牛的乳汁体细胞计数(白细胞和上皮细胞)上升,这些细菌主要传染部位为奶牛乳腺;环境型致病菌如大肠杆菌(*Escherichia coli*)和克雷伯杆菌难以在宿主体内长期存活,它们通过"入侵"宿主并在体内短暂繁殖,并诱导宿主产生免疫反应;机会型致病菌(如凝固酶阴性葡萄球菌)主要存在于乳腺皮肤及乳头部位,当奶牛机体免疫力正常时,这些病原体不会致病,而当机体免疫力下降时就会诱发感染。奶牛乳腺炎的预防、治疗和感染致病菌的种类紧密相关,多数情况下,传染型致病菌感染所引发的奶牛乳腺炎,通过优化管理策略,比如增强有效的挤奶和卫生措施的认知,能够限制革兰氏阳性细菌的传播,并使金黄色葡萄球菌分离株和亚临床乳腺炎在全球范围内的发生率显著降低。而其他微生物,例如大肠杆菌,能够导致奶牛在分娩前、

大肠杆菌感染致奶牛乳腺炎的致病机制及防治

后和泌乳早期患上乳腺炎,出现显著的局部,有时甚至是严重的全身临床症状,在最为严重的情形下,每年或许会导致数例奶牛死亡,而这一问题却较少受到关注。

一、大肠杆菌

乳腺炎是奶牛的常见疾病之一,具备全球性的经济影响。乳腺炎的症状体现为奶牛乳腺组织呈现红、热、肿、痛,主要因致病菌的侵入而致,从而使得牛奶产量和乳品质量下降。在中国以及其他地区的现代奶牛群体里,大肠杆菌已成为诱发奶牛乳腺炎的主要病原体之一。大肠杆菌作为一种环境病原体,于管理不佳且体细胞数偏高的奶牛场中,是产生临床型乳腺炎的常见致病根源。另外,大肠杆菌是奶牛急性乳腺炎的主要致病因子,通常恢复较为迅速;然而,在极端情形下,大肠杆菌性乳腺炎会引发严重的全身临床症状,诸如脓毒症并伴随发热;大肠杆菌感染还会导致亚临床和持续性的乳房感染。在英国,21%由大肠杆菌引发的临床乳腺炎病例中,均能观测到存在持续感染的乳区。因此,由大肠杆菌感染引起的临床型乳腺炎症状轻重不一,从轻度的局部症状到严重的临床症状均可能出现,甚至波及全身,最终造成感染动物的死亡。此外,大肠杆菌性乳腺炎病例的结果和严重程度主要取决于环境因素,以及奶牛针对病原菌相关分子模式(其中最为显著的是脂多糖)所产生的先天免疫反应。

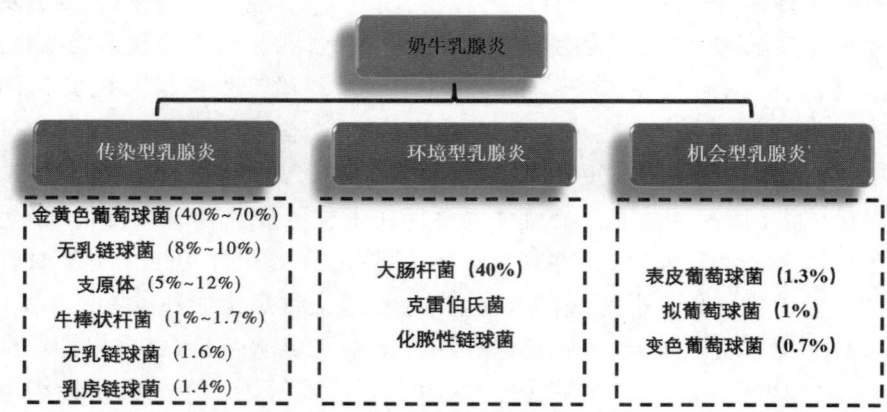

图 1-5 细菌性奶牛乳腺炎发病率

大肠杆菌又名大肠埃希氏菌，属于革兰氏阴性菌。大肠杆菌为兼性厌氧菌，在肠道正常菌群里属于优势菌种，对维持肠道生理机能有着关键作用。1885年，德国儿科医生Theobald Escherich首次将这种微生物界定为大肠杆菌，它是从正常婴儿粪便中分离出的一种短而粗壮的杆状菌。当时他觉得这是无害的腐生菌。在此后的半个多世纪，大肠杆菌一直被视作粪便中的主要共生菌，被认为是无毒的。然而，随着大量证据的涌现，表明大肠杆菌能够致使人类罹患疾病，这一观点逐步发生了转变。由致病性大肠杆菌引发的感染或许只局限于黏膜表面的定植，也或许会在全身扩散，并且涉及尿路感染、败血症、脑膜炎以及胃肠道感染等。

（一）大肠杆菌的生物学特性

依据抗原结构的差异，大肠杆菌能被划分为180多种血清型，镜检时呈现为单一或成对存在的短杆菌，两端呈钝圆状，无芽孢，周身带有鞭毛，具有微荚膜，属于好氧或兼性厌氧菌。在37℃以及pH 7.4～7.6的最适宜生存环境中能够展现其发酵能力。通常情况下，大肠杆菌对人和动物并无害处，不仅不会引发疾病，还能够维系肠道菌群的平衡，保障肠道的健康，然而，部分菌株可能通过获取致病基因从而转变为病原体。对10株大肠杆菌菌株的基因组DNA序列进行分析后发现，每个大肠杆菌分离株涵盖了4 000～5 000个基因，甚至更多，并且均共同拥有约3 000个相似的基因。因此，大肠杆菌能够被归类为不同的基因类型以及生物组。此外，大肠杆菌菌株还能够基于编码、荚膜多糖（K抗原）、鞭毛蛋白（H抗原）和脂多糖等众多传统遗传基因展开多样性分析，发现大肠杆菌拥有超过60种的H和K抗原，这显示出大肠杆菌菌株具有多样性。这种多样性的基因有助于大肠杆菌隐藏其毒力基因以及增强与其他共生菌的定植能力。大肠杆菌大多为无性繁殖，可被划分为6个主要的系统发育类群：A、B1、B2、D1、D2和E，也有将系统发育类群D1和D2分别命名为D和F，研究表明，引发乳腺炎的大肠杆菌菌株属于典型的共生菌，多数归属于系统群A。不过，也有研究指出引发乳腺炎的大肠杆菌菌株隶属于不同的系统类群。综上所述，大肠杆菌具备基因多样性、丰富的血清型以及系统发育多样性。

(二)大肠杆菌的分类

大肠杆菌是哺乳动物肠道微生物的共同构成部分,也是被研究得最为深入、操作最为广泛的微生物之一。大肠杆菌已被归类为至少6种不同的致病类型,涵盖肠致病性大肠杆菌、肠出血性大肠杆菌、产肠毒素大肠杆菌、肠聚集性大肠杆菌、肠侵袭性大肠杆菌和弥漫性黏附大肠杆菌。流行病学的研究成果表明,致使奶牛乳腺炎的大肠杆菌和其他大肠杆菌菌株相较,并未展现出特殊的毒力因子及血清型,大肠杆菌引发奶牛乳腺炎的毒力基础或许是所有大肠杆菌菌株共有的脂多糖、鞭毛等毒力因子。然而,不同的大肠杆菌性奶牛乳腺炎的临床表现却呈现出显著的差别,从温和且可自愈的乳腺炎到急性致死性乳腺炎。例如,大肠杆菌 P4 菌株具备很强的毒力特性,动物试验同样能够致使严重的脓毒血症奶牛乳腺炎,但其他大肠杆菌菌株在临床感染奶牛中则呈现出轻度乳腺炎症状或者处于潜伏状态。这种差异传统上被认为是源于宿主的免疫应答及基因组成的差异。然而,鉴于当下规模化养殖中奶牛的基因相近,而大肠杆菌临床分离株表现各异,因此很可能是奶牛乳腺炎的大肠杆菌携带和其他肠外致病菌不同的毒力基因,引发了一种全新的致病型——乳腺致病性大肠杆菌。

(三)大肠杆菌型奶牛乳腺炎

环境性乳腺炎在宿主和环境层面存在诸多潜在的致病要素与促成因素。在兽医实践当中,通常认为严重的临床型乳腺炎(乳汁和乳腺出现异常,伴有全身性体征)主要由大肠杆菌引发,会致使乳腺组织产生不可逆的损伤,彻底丧失产奶能力,甚至造成奶牛死亡。鉴于当下对牛场的管理愈发合理化以及对奶牛乳腺炎的重视程度提升,由大肠杆菌导致的奶牛乳腺炎也呈现出中度(乳汁和乳腺异常,无全身性体征)、轻度(仅乳汁异常)或者持续性亚临床(无显著体征)症状。还有研究报道称,大肠杆菌引起的乳腺炎是短暂性的,疾病的结局在很大程度上取决于宿主因素,涵盖是否处于哺乳期、能量平衡状况、维生素是否缺失以及疫苗接种状态。尽管大多数大肠杆菌感染是短暂的,然而对细菌菌株分子分型的纵向研究显示,大肠杆菌感染能够存在于乳腺中,反复发作的临床型乳腺炎与亚临床感染有关。亚临床大肠杆菌感染能够在干奶期起始,在哺乳早期呈现为临床型乳腺炎症状,大肠杆菌甚至能够在牛奶中持续存在长达100多天。为预防大肠杆菌性奶牛乳腺炎,首先要明确在哺乳早期临床型乳腺炎是由干奶期还是哺乳期的感染所导致的。需要针对不同时期的感染风险施行相应的控制举措,例如,在干奶期环境卫生不佳或不使用乳封剂,在哺乳期卫生或营养不足。针对这些风险因素,都应当尽量留意并加以避免。

(四)乳腺对乳腺致病性大肠杆菌的应答机制

致病菌如大肠杆菌进入乳头后,能够损坏乳房内主要的乳腺导管或导管的细胞壁,接着进入乳腺导管内进一步黏附和侵入奶牛乳腺上皮细胞,以免被乳汁冲刷掉,进而躲避宿主的免疫系统,随后大肠杆菌会迅速繁殖,使得大量的病原体侵入乳腺导管和乳腺肺泡中,并在乳腺组织中定植。大量定植于乳房内的大肠杆菌会促使乳房血管扩张,渗透性增强,致使大量免疫细胞从被感染的乳腺组织迁移至腺泡组织,导致体细胞计数升高,释放炎症介质,造成局部乳腺水肿。此外,部分致病性大肠杆菌产生的毒素会破坏乳腺细胞的完整性,随后打破血乳屏障,使血管中的凝血因子流向感染部位并与牛奶接触,导致乳汁凝固。乳腺内膜细胞的炎症反应致使乳腺导管内壁增厚,从而影响牛奶的流出,并阻塞乳导管,进而引发临床型乳腺炎。研究发现,50个菌落形成单位(CFU)就能够引起乳腺感染,激活先天免疫系统;随着乳腺内病原体增多,感染加剧,会引起持续或反复的乳房感染,进而导致乳腺上皮细胞坏死及凋亡等损伤(图1-6)。在轻度乳腺内感染的病例中,大肠杆菌会被宿主免疫系统规避和清除,因此发生炎症反应的血管和乳腺在数天内便会恢复正常;倘若感染持续的时间较长,就可能会导致乳腺导管阻塞,乳汁淤积会使乳腺组织处于静止状态,从而致使奶牛泌乳暂时停止或永久无乳。当损伤极为严重时,多数乳腺细胞遭到破坏,乳腺组织便开始萎缩,最终乳腺组织被结缔组织和坏死组织替代,给乳腺细胞带来不可逆转的损伤。

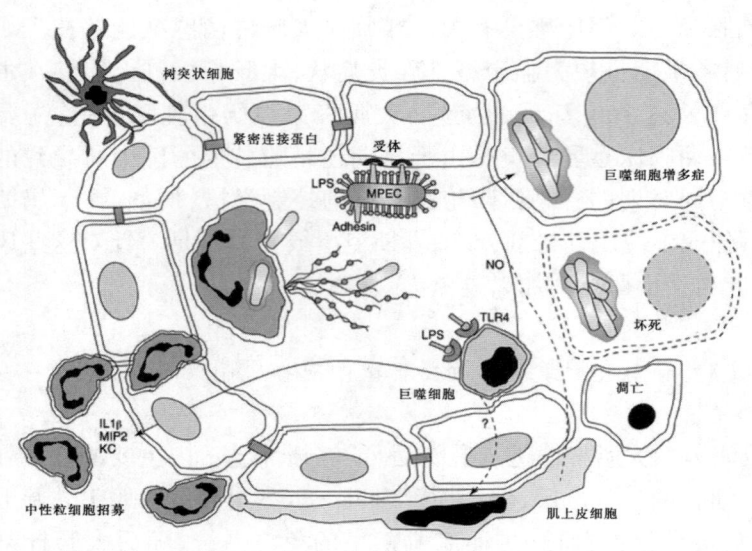

图 1-6　乳腺对乳腺致病性大肠杆菌的应答机制

二、金黄色葡萄球菌

（一）病原学

金黄色葡萄球菌（*Staphylococcus aureus*，简称金葡菌）属于革兰氏阳性菌，归为葡萄球菌属，又被称作"嗜肉菌"，能够引发人和动物的多种严重感染。此菌无芽孢、无鞭毛，多数情况下无荚膜。在液体培养物中，金葡菌镜检时呈现为单个、成对或者短链状排列；在固体培养基上，菌体镜检时则呈现出典型的葡萄串状排列。在普通培养基上，金黄色葡萄球菌生长状况良好，菌落表面光滑、隆起，圆形、湿润、不透明，且边缘整齐。其最适生长 pH 值是 7.4，营养需求不高，属于需氧或兼性厌氧菌。有些菌落呈现典型的黄色，有些则为白色。在血琼脂培养基（TSA）平板上生长的菌落较大，周边可见 α-、β-、双（α+β）溶血。金黄色葡萄球菌能通过甘露醇发酵试验阳性、过氧化氢酶试验阳性以及血浆凝固酶试验阳性的特点，与其他葡萄球菌进行鉴别。

（二）毒力因子

金黄色葡萄球菌的致病能力与其生成的多种毒力因子（如凝固酶、溶血素、肠毒素、耐热核酸酶等）存在关联。凝固酶由 coa 基因进行编码，由 670 个氨基酸构成，能够通过激活凝血酶原的作用使兔血浆发生凝集。研究显示，凝固酶的致病性依赖于金黄色葡萄球菌的存在。单纯给大鼠注射凝固酶时，大鼠的生存率未受影响；但当与凝固酶阴性的金黄色葡萄球菌一同注射时，大鼠的病死率会显著提高。

金黄色葡萄球菌产生的溶血素主要包括 α-溶血素、β-溶血素、δ-溶血素和 γ-溶血素，它们的致病机制各不相同。其中，α-溶血素是一种膜成孔毒素，由 hla 基因编码，约 33 kDa 大小，能够严重损伤红细胞、血小板，引发毛细血管平滑肌收缩痉挛，致使血流阻滞，造成局部缺血坏死。敲除 hla 基因后，菌株的毒力明显减弱。此外，在上皮细胞中，α-溶血素能使金黄色葡萄球菌免受囊泡胞吞作用。在巨噬细胞中，透射电镜表明金黄色葡萄球菌依赖 α-溶血素途径，通过破坏吞噬体膜结构来躲避吞噬细胞的吞噬作用。同时，宿主对 α-溶血素浓度的感应灵敏度会影响先天性免疫反应，也会对宿主针对该菌引发的全身性感染的防御能力产生影响。

迄今为止，共发现 21 种不同类型的肠毒素或类肠毒素。最为常见的金黄色葡萄球菌肠毒素是 SEA 和 SEB，二者均和食物中毒有关。SEA 是美国 80% 食物中毒事件的元凶，SEB 则占 10%。肠毒素蛋白具有显著的热稳定性、耐酸性以及耐胃肠蛋白酶灭活的特性。因此，即便经过高温处理，肠毒素蛋白的超抗原活性也不会丧失，并且能够轻易进入人体肠道并产生毒素。肠毒素的检测方式有动物实验（幼鼠和猿类）、免疫学方法（ELISA 商品试剂盒）、分子生物学方法（PCR 方法检测毒素基因、荧光定量反转录 PCR 技术研究毒素基因的相对表达）。金黄色葡萄球菌产生的肠毒素应当得到足够的重视，必须制定防治策略来应对其诱导的炎性免疫反应，尽可能从源头杜绝肠毒素的产生与传播。

生物膜形成与金黄色葡萄球菌体内感染过程中的致病特性密切相关，如持续耐药性、逃避细胞吞噬、促炎或抑炎等。细菌通过分泌细胞外多聚糖和蛋白质等成分，先构建起坚固的生物膜框架，然后更多的细菌黏附于其上，形成成熟的生物膜。Gogoi-Tiwari 研究了两株分离自临床型奶牛乳腺炎的金黄色葡萄球菌,发现其生物膜形成能力的强弱对鼠乳腺炎模型中乳腺组织的损伤程度有影响,结果表明,具有强生物膜形成能力的菌株诱导产生显著的急性乳腺损伤,表现为明显的坏死和中性粒细胞浸润;相反,弱生物膜形成能力的菌株引起的乳腺损伤程度明显减轻,尽管二者均能诱导产生炎性因子 IL-1β 和 TNF-α,但生物膜形成能力强的菌株诱导的 TNF-α 水平明显更高。研究显示,除了 ica 基因调控的 PIA 外,其他非 ica 依赖途径如 agr(附属基因调节子)操纵子、sarA(葡萄球菌附属调节子 A)基因、sasG(金黄色葡萄球菌表面蛋白 G)基因、bap(生物膜相关蛋白)基因和 ccpA(分解代谢控制蛋白 A)基因也对生物膜的形成具有重要的影响。目前,存在多种用于检测生物膜形成的方法,如传统的刚果红培养基法、改良微孔板法和激光共聚焦显微镜法。生物膜形成与金黄色葡萄球菌耐药性的产生显著相关,可能的机制为:其一,降低了抗菌药物的渗透性;其二,吸附钝化酶,水解抗菌药物;其三,细菌代谢水平低,对抗菌药物的敏感度下降。

(三)金黄色葡萄球菌型奶牛乳腺炎

金黄色葡萄球菌在全球范围内是牛乳腺内感中最为常见的分离病原体之一,也是从原料奶中分离出的最常见的传染性乳腺炎病原体,它能够对所有的抗生素产生耐药性,这种细菌在很大程度上能够耐受不利于自身生长的环境。金黄色葡萄球菌的耐受能力与其多样化的基因功能有关,例如产生耐甲氧西林的金黄色葡萄球菌菌株,在宿主上形成生物被膜,这些都增强了金黄色葡萄球菌在乳腺炎疾病中的致病能力,使其能够躲避宿主的免疫系统并产生多重耐药。

金黄色葡萄球菌同样存在于牛场环境当中,奶牛可能会被来自环境中的金黄色葡萄球菌所感染,而非源自其他奶牛。研究发现,100% 的加州奶牛群和 93% 的英国奶牛群中均存在金黄色葡萄球菌感染。在英国,感染金黄色葡萄球菌的奶牛比例约为 21%,丹麦则为 10%。长期的调查表明,金黄色葡萄球菌对于乳制品行业的重要性始终未变。金黄色葡萄球菌可在多个方面造成经济损失:牛奶产量降低,牛奶被细菌污染以及牛奶体细胞数增多,兽医和治疗费用增加,还有奶牛被过早淘汰等。牛奶样本中金黄色葡萄球菌大量生长的奶牛被扑杀的风险颇高。

在中国,金黄色葡萄球菌性奶牛乳腺炎的发生率高于其他国家,治愈的难度以及牛场中慢性乳腺炎的存在导致了高感染率。金黄色葡萄球菌的耐药性伴随抗生素的广泛运用而出现,耐药性的产生是乳腺炎低治愈率的主要原因。此外,更为重要的是,金黄色葡萄球菌菌株多重耐药的产生,对于乳腺炎的治疗是一个巨大的挑战,并且具有一定的公共卫生意义。

耐甲氧西林的金黄色葡萄球在人医领域是极为重要的病原体,近些年来在奶牛方面也是如此。耐甲西林金黄色葡萄球菌在人医和兽医方面均有越来越多的相关报道。动物源性耐甲氧西林金黄色葡萄球菌或许会导致人类的感染。有报道指出,对于人类感染的耐甲氧西林的金黄色葡萄球菌,动物或许充当着传播媒介的角色。在治疗和防控奶牛乳腺炎时,研究耐甲氧西林金黄色葡萄球菌的耐药机理和流行病学是极为重要的。

三、肺炎克雷伯杆菌

(一)病原学

克雷伯氏菌属(Klebsiella)属于无运动力但有荚膜的革兰氏阴性杆菌,共分为5个种:肺炎克雷伯氏菌、产酸克雷伯氏菌、植生克雷伯氏菌、土生克雷伯氏菌和变形克雷伯氏菌。

在克雷伯氏菌属当中,肺炎克雷伯杆菌(*Klebsiella pneumonia*)对人畜的致病性居于首位,其所导致的疾病占据该菌属感染的95%以上,所以当下的研究也主要聚焦于肺炎克雷伯杆菌。这种菌分为鼻硬结亚种、肺炎亚种和鼻炎亚种。肺炎亚种的肺炎克雷伯氏菌,又被称作肺炎杆菌,能够引发奶牛乳腺炎。1882年,Friedlander首次从大叶性肺炎患者的痰液里分离出肺炎克雷伯杆菌。该菌在自然界分布广泛,致病性强,是一种人畜共患病病原菌,能够感染人体和动物的全身,尤其是呼吸和泌尿系统,致使人和多种动物患上肝脓肿、肺炎和脑膜炎等病症,败血症等严重疾病也有可能由此引发。

（二）形态特征与培养特性

肺炎克雷伯杆菌是一种短粗杆状、卵圆形的革兰氏阴性杆菌，常常呈现单个、双排或者短链状的排列形式。肺炎克雷伯杆菌的大小为（0.3~1.0）μm×（0.6~6）μm，具有显著的荚膜结构，无鞭毛，有时在少部分菌体中可能会出现膨大自体溶解的现象。在观察肺炎克雷伯杆菌的荚膜时，由于细菌荚膜对光的反射率低，通常采用诸如湿印度墨汁遮盖法之类的负染色法来替代常规染色方法，以观察其荚膜结构。肺炎克雷伯杆菌在pH为5.5~10均可生长，其最适宜的pH是7.0~7.6。该菌能够生存的环境温度在15~45℃，最佳生长温度为37℃，超过50℃该菌就会被灭活。肺炎克雷伯杆菌依据代谢类型可分为发酵型和呼吸型，属于兼性厌氧菌。该菌对营养的要求相对较低，在普通培养基上便能生长。在绵羊血—大豆酪蛋白琼脂培养基上会形成较为黏腻的、淡灰白色的光滑圆形菌落，无溶血迹象。在麦康凯固体培养基上培养24 h后，菌落呈现为湿润的淡粉红色。在伊红美蓝固体培养基上培养较长时间后会形成凸起的黏性菌落，中间呈黑灰色，外周为半透明红色，且有扩散的趋势，有时会聚集在一起形成一片菌苔样。在强选择性琼脂培养基上也会形成类似于在麦康凯固体培养基上生长的粉红色黏性菌落，并且在强选择性琼脂培养基上生长48 h后，菌落和培养基都会变为黄色。此外，为了从大肠杆菌污染料中分离鉴定肺炎克雷伯杆菌，人们研发了多种高效选择性培养基。常用的有麦康凯—肌醇—羧苄青霉素琼脂培养基和西蒙氏柠檬酸盐固体培养基。这两种高效选择性培养基的筛选原理主要是：肺炎克雷伯杆菌能够利用西蒙氏柠檬酸盐固体培养基中的柠檬酸盐里的碳源为自身生长提供能量，形成蓝色的菌落。另外，肺炎克雷伯杆菌在麦康凯—肌醇—羧苄青霉素琼脂培养基上呈现出红色圆形菌落的特征，肺炎克雷伯杆菌可以利用该培养基中的肌醇作为发酵物质，并且杂菌的生长会被培养基中所含的羧苄青霉素所抑制，从而便于肺炎克雷伯杆菌的筛选。

（三）肺炎克雷伯杆菌的毒力因子及毒力基因

肺炎克雷伯杆菌表面多糖存在高黏表型，rmpA编码高黏液性表型调节因子A，MagA阳性的菌株具备较强的抗血清和抗吞噬作用，并与K1型肺炎克雷伯杆菌相关。依据荚膜抗原（K抗原）的不同，肺炎克雷伯杆菌可分成82个血清型，其中血清型K1、K2、K5会致使人和动物患上严重的感染性疾病，K54和K57型的菌株可引起群体侵入性肝脓肿综合征。高黏性肺炎克雷伯杆菌的荚膜类型涵盖K1、K2、K5、K16、K20、K54和K57。肺炎克雷伯杆菌所具有的脂多糖能够激活补体，使C3b沉积于脂多糖上，进而阻碍溶膜体复合物的形成。$wcaG$靶基因能够编码合成荚膜盐藻糖，增强细菌逃避巨噬细胞吞噬的能力，$wabG$能够编码分泌细胞黏附式荚膜多糖。

肺炎克雷伯杆菌具备黏附因子，这些因子主要包含菌毛和非菌毛黏附蛋白。菌体借助菌毛与宿主细胞充分接触或者黏附，肺炎克雷伯杆菌有Ⅰ型和Ⅲ型菌毛，Ⅰ型菌毛的毒力基因$fimH$能够编码菌毛尖端黏附素；mrkD和mrkA编码Ⅲ型菌毛，促进生物膜的形成。非菌毛黏附蛋白是细菌表面并非菌毛形态的各种具有黏附作用的结构物的统称，主要存在于革兰氏阳性菌中，在革兰氏阴性菌中较为少见。肺炎克雷伯杆菌能够分泌铁载体蛋白，且竞争宿主体内铁离子，铁载体蛋白有肠菌素和产气菌素两大类。ntB、ybtS、iutA、kfuBC编码产物是铁摄取系统的重要成分；$alls$靶基因编码尿囊素调节基因的活化剂，并分泌产气菌素；毒力基因也和铁摄取系统有关。

（四）肺炎克雷伯杆菌型乳腺炎

在临床乳腺炎病例当中，约40%的病原菌属于革兰氏阴性菌。肺炎克雷伯杆菌是除大肠杆菌外致使奶牛乳腺炎的极为重要的革兰氏阴性菌。肺炎克雷伯杆菌侵入乳腺之后能够黏附在乳腺上皮细胞上，侵入胞内并保持自身活力，致使乳汁冲刷作用减弱，减少了被免疫细胞吞噬的机会。所以该菌对乳腺组织具有很强的致病力，其引发的乳腺炎症状严重、治疗困难，奶牛容易被淘汰或者死亡。2014年至2017年间，本项目组的高健等人收集了来自21个省161家农场的3288份临床乳腺炎奶样，经过分离鉴定，发现克雷伯杆菌的分离率为13.0%，这一比例高于金黄色葡萄球菌、停乳链球菌、无乳链球菌等常见致病菌，表明其在我国已成为引发奶牛乳腺炎的主要致病菌。研究还发现奶牛也可能通过接触而感染肺炎克雷伯杆菌，这种类型的乳腺炎会使牛奶产量大幅降低，奶牛死亡率上升,从而造成严重的经济损失。奶牛乳腺感染肺炎克雷伯杆菌后,会出现严重的炎症反应和组织坏死,产奶量下降,乳房变得坚硬,牛乳性状改变;同时,奶牛的全身症状明显,发热、精神萎靡、食欲减退;而且,抗生素的治疗效果不显著。

处于干奶期的奶牛感染肺炎克雷伯杆菌的风险较大。在干奶期,尤其是干奶开始后的两周以及产犊前的两周,乳房对肺炎克雷伯杆菌的敏感性较强,也就是说患病的风险较高。依据牛场的数据统计,肺炎克雷伯杆菌乳腺炎主要发生在新产牛阶段。这主要是因为各种环境性病原菌在干奶期进入到乳房内。此外,在干奶期间肺炎克雷伯杆菌更易于在乳腺内生存与繁殖,由于处于干奶期间的奶牛无法分泌乳汁,也就缺少了乳汁中乳铁蛋白的抑菌作用,为肺炎克雷伯杆菌的侵入以及生长繁殖创造了条件。同样,泌乳期的奶牛也存在一定感染肺炎克雷伯杆菌的风险。但有所不同的是,泌乳期内母牛感染的平均时间相对较短,通常为21天左右,很少有超过90天的持续性感染。尽管有研究表明细菌感染与体细胞数没有显著的相关性,但是感染细菌的奶牛的乳汁中也可能具有较低的体细胞数且不呈现出明显的症状。然而,持续时间较短的感染很容易导致体细胞数升高,从而使原料奶无法通过质量检测或者降低其品质。肺炎克雷伯杆菌性乳腺炎的发病情况具有一定的季节性。炎热的夏季是肺炎克雷伯杆菌性乳腺炎的高发季节,一方面夏季多雨闷热且潮湿,有利于肺炎克雷伯杆菌的生存;另一方面,高温容易引起奶牛的热应激,从而使奶牛对克雷伯杆菌的敏感性提高,更易被克雷伯杆菌侵入并感染。然而,并非所有地区都是这样的状况。美国的研究者 Paulin 等在不同季节从印第安纳、明尼苏达、威斯康星和宾夕法尼亚 4 个州的 6 个养牛场收集并分析了 635 份乳样,结果显示,克雷伯杆菌在冬季检测出的比例为 36.2%,夏季为 32.6%,春季和秋季比例较低,为 15.6%。而后,其又在不同季节从以上牛场内收集了 93 个环境样本(主要为沙床),并从中分离出 31 株克雷伯杆菌,其中春季和冬季分离出克雷伯杆菌的比例为 71%(22/31),夏季占比为 19.4%(6/31),秋季占比为 9.7%(3/31)。这些数据表明,在以上地区,肺炎克雷伯杆菌在冬季环境中存在较多,并且奶牛也最容易在冬季患上肺炎克雷伯杆菌乳腺炎。

四、无绿藻

（一）病原学

1894 年 Wilhelm Kruger 首次分离出左氏无绿藻之后，有关左氏无绿藻致使人类、犬和奶牛等动物患病的文献报道日益增多。无绿藻（*Prototheca spp*）是一种不含叶绿素的真菌样单细胞绿藻。无绿藻这类生物自身不含胞壁酸及真菌特有的葡糖胺，又因缺失叶绿素和具有不同于藻类的细胞壁（在电镜下细胞壁为二层，而非三层）而被归类为无绿藻属。此外，无绿藻依靠无性繁殖生成内孢子，在适宜的营养条件下，每 5~6h 增殖一代，其形态呈球形到椭圆形或楔形。1894 年 Kruger 确定了无绿藻的 6 个属，涵盖莫氏无绿藻（1894 年发现）、斯氏无绿藻（1985 年发现）、尤氏无绿藻（1987 年发现）、韦氏无绿藻（1959 年发现）、左氏无绿藻（1894 年发现）和伯氏无绿藻（2006 年发现）。其中，左氏无绿藻、伯氏无绿藻和韦氏无绿藻被证实对人类和动物具有致病性，而左氏无绿藻是无绿藻乳腺炎中最为主要和常见的病原菌。

当下，无绿藻主要包含大型无绿藻（*Prototheca stagnora*）、中型无绿藻（*Prototheca zopfii*）、小型无绿藻（*Prototheca wickerhamii*）以及微型无绿藻（*Prototheca cutis*）。其中，依据基因型的差异，中型无绿藻分为希弗氏无绿藻（*Prototheca ciferrii*，之前称作无绿藻 GT-Ⅰ）和牛无绿藻（*Prototheca bovis*，之前称为无绿藻 GT-Ⅱ）。但目前基于无绿藻线粒体 CytB 基因提出了一种新的分类方式（图 1-7），包含与奶牛有关的 *P. ciferrii* 和 *P. blaschkeae*；与人类有关的 *P. wickerhamii*，*P. cutis* 和 *P. miyajii*；其余的为 *P. moriformis*，*P. stagnora*，*P. tumulicola*，*P. zopfii*，*P. cookei*，*P. pringsheimii*，*P. xanthoriae* 和 *P. cerasi*。在与奶牛相关的物种中，*P. ciferrii* 在奶牛的粪便和牛场的环境里被发现，而 *P. blaschkeae* 在奶牛的粪便、牛场的环境以及乳腺炎临床样品中均有发现，并被全球公认是引发奶牛群临床无绿藻性乳腺炎的主要病原菌。

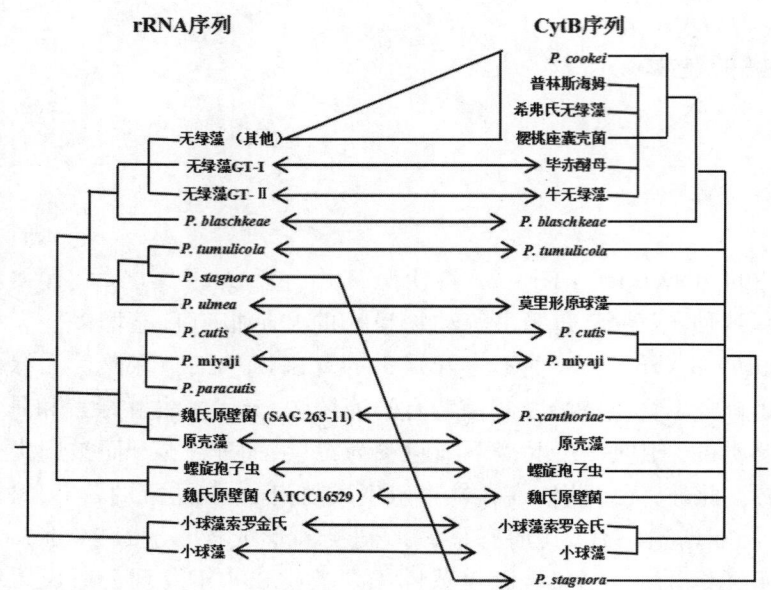

图 1-7　原膜菌/螺旋孢子虫/小球藻/小球藻谱系内物种/基因型之间的关系

　　1952年,左氏无绿藻首次被认定为奶牛乳腺炎的一种病原体,导致产奶量大幅下降,出现白色凝乳块的水样分泌物。依据18S rDNA的序列分析,左氏无绿藻被划分为3个亚型:基因1型、基因2型和基因3型。生化鉴定能够精准地将这三种基因型区分开来。而运用傅里叶变换红外光谱法(FTIR),可以进一步观察到基因3型与其他两个亚型之间更为显著的差异,然而,该方法却无法区分基因1型和基因2型之间的差异。此外,这三种基因型,每一种通过免疫印记都能生成特异性抗体形式。Roesler等依据3种不同基因亚型的18S rDNA构建系统发育树,从而明确了3种亚型之间的差别,因此,基因3型被重新确定为一个新的无绿藻亚种即伯式无绿藻,而左氏无绿藻则正式被分为左氏无绿藻基因1型和左氏无绿藻基因2型。

　　无绿藻作为奶牛乳腺炎的致病菌,长期以来一直被忽略,因为这类病原菌常常被误认为是酵母菌或污染腐生菌。在分离奶牛乳腺炎奶样中的这一病原菌时,通常采用沙氏葡萄糖琼脂培养基进行培养,往往培养的时长要多于其他细菌才能够形成较为明显的菌落。在菌落形态特征方面,无绿藻和 P. ciferrii 的菌落均呈现出典型的形态学特征,无绿藻的菌落在沙氏葡萄糖琼脂培养基呈灰白色,表面中央突起呈粒状或锯齿状,而 P. ciferrii 的菌落呈乳白色,表面光滑隆起;在电子显微镜下,无绿藻具有孢子囊的典型特征,从球形到椭圆形,包含孢子囊孢子,而 P. ciferrii 显示出紧凑、圆形,大小约为无绿藻的两倍。此外,对于这两种菌的鉴定,在生化方面,无绿藻比 P. ciferrii 具有更多的二十碳二烯酸,在分子特征方面,通常使用基因型特异性 PCR 测定 18S rDNA 序列并结合 CytB 基因对其进行鉴定。由此可见,无绿藻和 P. ciferrii 之间在菌落形态方面存在差异,在生化和分子特征方面也各不相同。

(二)无绿藻的致病性

无绿藻广泛存在于污水、土壤、粪便、河流、湖泊等环境当中,这种藻类对于环境条件具备很强的适应性,尤其是高湿度以及富含有机物的环境。虽然在山羊、马和非驯养动物里有零星的病例报告,但牛、狗和猫是分离出无绿藻物种的主要家畜。在临床上,由无绿藻感染所引发的奶牛乳腺炎通常面临难以鉴定、难以治疗等问题,给牧场造成巨大的经济损失。无绿藻感染奶牛乳腺后会致使乳房外观发生改变(如肿胀、发红),同时引起乳汁品质变化,并且该菌对大多数抗生素具有严重的耐药性,运用抗生素进行治疗时效果欠佳。体外研究显示,无绿藻侵染奶牛乳腺上皮细胞后,能够大量地黏附在细胞表面,降低细胞活性,诱导细胞凋亡,而 P. ciferrii 对细胞造成的损伤明显低于无绿藻。此外,无绿藻持续性感染奶牛乳腺上皮细胞后能够显著增加胞内氧化应激的代谢产物丙二醛(MDA)的含量,并促进胞内活性氧(ROS)的蓄积,同时降低胞内调节氧化应激相关酶的活性,比如超氧化物歧化酶(SOD),造成胞内氧化与抗氧化系统失衡,最终导致氧化应激。在无绿藻诱导的小鼠乳腺炎模型中,无绿藻感染后破坏了乳腺组织形态结构,导致大量的炎性细胞浸润,同时上调了小鼠乳腺上皮细胞内凋亡相关基因 Bax 和 caspase-3 的表达,致使乳腺上皮细胞发生凋亡;此外,无绿藻感染能够使宿主细胞炎性因子分泌量上升,引发剧烈的炎性反应,虽然无绿藻感染诱导奶牛乳腺上皮细胞炎症反应,同时又诱导凋亡,但在感染 12 h 内炎症占据主导地位,随着感染时间的延长,细胞的凋亡率逐渐升高。无绿藻作为一种机会性致病菌,能够导致奶牛乳腺炎、犬猫皮肤病以及人类的皮肤疾病。通常情况下抗真菌药物的治疗效果不佳,而人源抗菌肽 LL-37 对无绿藻具有杀灭作用,LL-37 能够降低无绿藻感染的小鼠巨噬细胞和乳腺上皮细胞中 TNF-α、Cxcl-1 和

IL-1β 的含量，降低炎症反应。

在临床上，无绿藻对于伴侣动物犬和猫也具有一定的致病性，伤口污染和创伤性皮下损伤是伴侣动物感染无绿藻的主要途径。在伴侣动物免疫力低下时，经伤口侵入的无绿藻可引发全身性或系统性感染。由于该菌在自然界中广泛存在，某些职业群体如稻田工人、渔民等也容易感染该菌，常见的感染部位或系统涵盖皮肤、鹰嘴囊、呼吸道和消化系统等，通常引发皮肤病、鹰嘴滑囊炎等。由此可见，该菌的致病性较强且宿主范围广泛。

(三) 无绿藻型奶牛乳腺炎

无绿藻在奶牛场中普遍存在，1952 年，无绿藻首次被确认为引发奶牛乳腺炎的病原菌。该菌感染奶牛乳腺后通常致使产奶量急剧下滑和乳房外观改变，同时致使乳汁品质发生变化，容易成为感染源头，最佳的控制办法就是淘汰患病牛。由无绿藻感染导致的奶牛乳腺炎通常呈现为慢性感染，但在某些情形下能够观察到具有临床症状的急性形式。无绿藻感染引发的奶牛乳腺炎在全球范围内广泛传播，其中，在比利时、意大利、德国和加拿大等国家均有病例出现。国外牧场的一项针对大罐奶中病原菌的检测表明，从大罐奶中分离获得 276 株病原菌，通过种属鉴定发现酵母 184 株，其中无绿藻占 92 株，且这一病原菌在同一牧场中被反复检测到。据报道，高健等人对采集的 620 份奶牛乳腺炎奶样和 410 份环境样品（包括粪便、饲料、垫料和饮用水样品）中的病原菌进行分离培养鉴定，从乳腺炎奶样中获取 84 株无绿藻，从环境样品中分离出 21 株无绿藻，而且这种类型的无绿藻性乳腺炎已在北京、天津和山东等地出现。无绿藻作为一种环境性致病菌，其传播方式也各不相同。通常状况下，无绿藻感染奶牛乳腺后，导致乳汁中携带无绿藻，乳汁中的无绿藻可被犊牛摄入并排泄到环境中；此外，受污染的牛奶还可能致使挤奶机械上存在无绿藻，这也是一种更为直接的传播途径（图 1-8）。由此可见，无绿藻作为引发奶牛乳腺炎的致病菌，其临床感染率逐年递增，而且无绿藻感染奶牛后不仅容易造成巨大的经济损失，还对食品公共安全存在潜在的威胁。

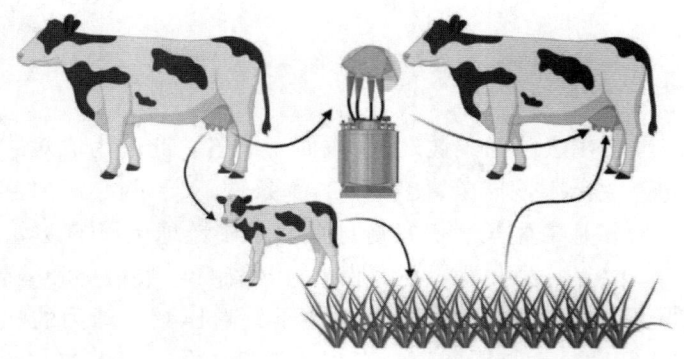

图 1-8 无绿藻在成年奶牛—犊牛—环境之间的传播

五、链球菌

链球菌(*Streptococci*)归属于链球菌属(*Genus streptococcus*),革兰氏阳性,菌体呈球形或卵圆形。其粒径为 0.6~1.0 μm,多为链状排列,短的链由 4~8 个菌体构成,长的链则由 20~30 个菌体组成。链长与细菌种类及生长环境紧密相关,通常在液态环境下的链长大于固态环境下的链长。该菌株无孢子,无鞭毛,有些菌株能在液相中形成荚膜,并呈链状结构。该菌属在自然界广泛分布,可在水、大气、灰尘、人类及动物的鼻咽、肠道等环境中被发现。D 群链球菌可分为肠球菌型和非肠球菌型。牛链球菌属于非肠球菌型 D 群链球菌,存在于 5%~16% 的人类肠道中,在人类病患中,它常作为肠道菌群的一部分被检测到,此外,在慢性肝病和胆囊疾病中也有所描述。2002 年,Poyartd 等描述了链球菌的两个新的亚种,分别是巴黎链球菌(*S. lutetiensis*)和巴氏链球菌(*S. pasteurianus*),并指出其在人类和动物之间能够共生。此后,基于甘露醇发酵和 β-葡萄糖醛酸酶活性,将牛链球菌分为 3 个生物型:(1) Ⅰ 型包括 *Streptococcus gallolyticus* ssp. *gallolyticus*;(2) Ⅱ/1 型包括 *S. lutetiensis* 和 *S. infantarius*;(3) Ⅱ/2 型包括 *S. galloly-ticus* ssp. *pasteurianus*。1945 年,有关牛链球菌群引起人类疾病的报道指出,巴黎链球菌(*S. lutetiensis*)、婴儿链球菌(*S. infantarius*)以及解没食子酸链球菌巴氏亚种(*S. galloilyticus* subsp. *Pasteurianus*)与非结肠肿瘤和脑膜炎关系密切。

(一)巴黎链球菌

巴黎链球菌在法国巴黎被发现,故而得此名。此菌为成对或以短链形式呈现的革兰氏阳性球菌,能够在37℃条件下,于脑心浸出液肉汤和葡萄糖肉汤培养基中生长。巴黎链球菌属于牛链球菌属D群链球菌分类下的Ⅱ/1型,会引发人类的心内膜炎、尿路感染、败血症和脑膜炎,且容易导致男性和老年人出现菌血症。当下针对巴黎链球菌的病例数量较少,仅在儿童和小动物胃肠道内有相关报道。金东等人在儿童的肠道中分离出了巴黎链球菌,并在巴黎链球菌基因组中总计发现20个基因组岛,涵盖5个耐药性基因组岛和2个毒力岛。基因组岛18编码黏附定植相关的因子基因,存在8个毒力因子,包括与纤连蛋白结合相关的pavA,与增殖相关的Lmb、slrA和rrgA,与上皮细胞黏附相关的srtA和psaA,与纤溶酶原结合相关的Eno,与先天性免疫逃避相关的cppA,与链球菌溶血素S的生物发生相关的htrA/degP,与抵御应激和形成生物膜相关的tig/ropA。在一位患有胆囊炎的患者血液中,分离出携带*lnuB*基因的巴黎链球菌,*lnuB*基因编码林可酰胺核苷酸转移酶,该酶在DNA的合成和修复过程中发挥着极为重要的作用,对细胞的增殖和分化意义重大。Piva等在2019年报道,从一只患有肠淋巴瘤的猫身上分离出一株耐多药的巴黎链球菌,该菌对恩诺沙星、克林霉素、马波沙星和四环素表现出耐药性。

(二)停乳链球菌

停乳链球菌属于兰氏C群链球菌,在血平板上呈现α溶血或不溶血现象,通过显微镜观察为链状球菌。停乳链球菌分为2个亚种:停乳链球菌似马亚种(*Streptococcus dysgalactiae* subsp. *equisimilis*)和停乳链球菌停乳亚种(*Streptococcus dysgalactiae* subsp. *dysgalactiae*)。前者常从人和动物中分离得出,是导致侵入性感染的重要人类病原菌,涵盖坏死性筋膜炎和链球菌中毒性休克综合征,而后者通常从动物中分离,例如马、牛、鱼和猪。值得注意的是,停乳链球菌停乳亚种分离株具有很强的细胞侵袭性,可用于体外侵袭损伤人类原代角质形成细胞,这表明停乳链球菌停乳亚种具备感染人体组织并引发疾病的潜力。因此,牛乳中停乳链球菌停乳亚种的高流行率或许能够使其成为一种食源性病原菌,对人类健康构成威胁。2个亚种生化特性的主要区别在于：停乳亚种水杨苷利用呈阴性,似马亚种水杨苷利用为阳性；停乳亚种一般为α溶血,似马亚种为β溶血。在显微镜下,停乳链球菌呈圆形或卵圆形,直径在0.5～1.0 μm。通常情况下,两个或者多个排列成行,没有荚膜。属于革兰氏阳性菌。该菌无法在普通培养基上进行培养。在血琼脂平板上培养时,其菌落呈圆形,表面光滑,有凸起和β溶血环。将其置于具有血清的肉汤培养基内,最初液体呈乳状,之后,在试管底部能够发现沉淀,而其上清澄清。触酶试验、CAMP试验、马尿酸钠水解试验、甘露醇、山梨醇、七叶苷、菊糖、过氧化氨酶和6.5% NaCl实验均为阴性,β葡萄糖醛酸苷呈阳性,能够对乳糖和海藻糖进行分解。

  停乳链球菌的致病性源于其能够产生多种毒力因子,包含黏附因子、酶、免疫逃逸因子、免疫反应性抗原、铁摄取系统、锰摄取系统、蛋白酶、超抗原、毒素等。这些毒力基因促进停乳链球菌停乳亚种与宿主血浆和细胞外蛋白的相互作用,黏附于上皮细胞,随后内化并传播至宿主组织。Gap C基因是编码Gap C蛋白的毒力基因。Gap C蛋白是近年来研究较多的链球菌毒力因子之一,它位于细菌表面,具有甘油醛-3-磷酸脱氢酶活性,该酶能够可逆地催化3-磷酸甘油醛转化为1,3-二磷酸甘油酸；另外它与纤溶酶原、肌动蛋白、纤维蛋白原等结合,也能够诱导B细胞增殖与分化,在疾病发展过程中发挥重要作用。Perez-Casal等研究证实,停乳链球菌、无乳链球菌和乳房链球菌在基因和蛋白质水平上均具有很高的同源性。Fontaine等利用乳房链球菌Gap C蛋白免疫奶牛,有良好的免疫保护性。*cyl*基因是编码溶血素的基因,通常位于质粒或染色体的毒力岛上,以操纵子的形式存在,由cylA、B、L、M、I等组分丛集而成,8个开放阅读框（cylR1、cylR2、cylLL、cylLS、cylM、cylB、cylA及cylI）依次排列,均参与溶血素的表达调控。溶血素是由细菌分泌的一种穿孔毒素,它能够通过溶解细胞膜、暴露细胞质来致使细胞死亡。溶血素的毒力作用不仅体现在溶血特性上,还表现在对上皮和内皮细胞的黏附和侵袭能力以及对宿主细胞炎性因子的激活能力上,因此被认为可能是主要诱导奶牛乳腺炎的毒力因子之一。MIG基因是编码MIG蛋白的毒力基因。Jonsson H等在停乳链球菌细胞表面鉴别到一种蛋白受体,因其具有能同时结合α2-巨球蛋白和IgG的特点,所以被命名为MIG蛋白。Song等人的研究发现停乳链球菌不仅能够通过

MIG 蛋白结合 IgG 和 α2-巨球蛋白来增强对中性粒细胞的抵抗力,而且可以通过 MIG 蛋白结合 IgA 来突破乳腺黏膜免疫系统的能力,因此认为 MIG 蛋白可能是停乳链球菌诱导奶牛乳腺炎的主要毒力因子之一。

韩博教授实验室在 2014 年至 2016 年对中国 160 余个大型牧场的临床乳腺炎样品进行了细菌培养和分离鉴定,停乳链球菌的分离率为 10.5%,是流行率最高的革兰氏阳性球菌,这证实停乳链球菌是最为重要的乳腺炎病原菌之一。Bi 等人关于牛场大罐奶中细菌流行情况的调查也显示,在 894 个牛场的大罐奶样品中,72.3% 的样品为停乳链球菌感染,说明停乳链球菌可能引发大量的隐性乳腺炎发病。在奶牛中,停乳链球菌会导致持续性感染的临床和亚临床型乳腺炎。由停乳链球菌引起的乳腺炎通常发生在泌乳初期,造成的产奶量损失可达每天每只牛 1.2～2.0 kg。停乳链球菌是中国、加拿大、德国、芬兰、法国、澳大利亚和美国等多个国家奶牛乳腺炎的主要病原菌之一。

(三)无乳链球菌

无乳链球菌(*Streptococcus agalactia*)在奶牛中属于一种严格专性乳腺内寄生且具传染性的病原体,最早于 1930 年由 Rebecca Lancefield 从牛乳腺中严重分离得出。无乳链球菌属于革兰氏阳性球菌,归属于 B 群链球菌属,依据菌株表面所带的特异性抗原,可将其划分为 10 个不同的血清型,是乳腺炎的主要传染性致病菌之一。无乳链球菌对营养的要求较为苛刻,在普通细菌培养基中生长不佳,在添加有 5% 绵羊血琼脂平板上生长良好。在 37℃培养 18～24 h 后,可见呈圆形、表面光滑、灰白色、直径为 0.1～1.0 mm 的小菌落。无乳链球菌的一些菌株可产生黄色或砖红色色素,多数菌株会形成溶血环,部分菌株不溶血,且溶血环的大小类型因菌株而异。无乳链球菌无鞭毛,不运动,不形成芽孢,有些能够形成荚膜,在含有 2% 血清的肉汤培养基中,生长初期呈现均匀浑浊,后期上部澄清,在管底可见絮状沉淀。镜检时常为单个、成双或长短不一的链状排列的革兰氏阳性球菌。其生化特性为:触酶阴性,海藻糖分解、马尿酸钠、CAMP 试验和 VP 试验为阳性,而七叶苷试验、6.5% NaCl 试验、山梨醇为阴性试验。常用七叶苷试验、CAMP 试验等方法与其他的乳腺炎致病性链球菌(乳房链球菌、停乳链球菌等)进行区分。

无乳链球菌在分离到的乳腺炎链球菌属中占据很大的比例,对奶牛的健康以及经济效益的影响愈发显著,受到了众多学者的关注。黄瑛在对采集的105份样品进行病原菌分离鉴定时,共分离到无乳链球菌58株,占分离细菌总数的76.32%,殷波从奶样中分离到的7种病原菌,其中无乳链球菌占40.0%。袁永隆等人对采集的来自23个省、市、自治区40个牧场的3006份奶样进行细菌学调查,结果显示主要的致病菌为无乳链球菌和金黄色葡萄球菌。李宏胜等人从我国30个城市采集2858份奶样,经细菌分离鉴定后发现,主要病原菌为无乳链球菌,所占比率为38.47%,停乳链球菌和金黄色葡萄球菌在分离中所占比例较少,分别为28.05%、17.94%。有些学者报道,部分地区的无乳链球菌的感染率已经超过了金黄色葡萄球菌,成为影响奶牛乳房健康的主要病原菌,有的地区无乳链球菌的检出率甚至高达70%,这一比例显著高于国外的平均检出率。近年来,国外有报道称在牛乳及其乳制品中也检测到了无乳链球菌,占比高达63.9%。在欧洲和北美的大部分国家,成熟的乳腺炎控制体系使奶牛无乳链球菌的患病率稳定维持在较低水平,普遍小于10%。然而,无乳链球菌依然对许多国家构成严重威胁。尤其是在那些乳品行业刚刚起步发展的国家,如哥伦比亚和巴西,在畜群中无乳链球菌流行率为40%~60%。牛群中无乳链球菌流行率较高的国家还包括西班牙(36%)、德国(29%)和中国(92%)。一些已经成功控制无乳链球菌的国家,如丹麦和挪威,仍会反复出现无乳链球菌的感染和流行。无乳链球菌的反复出现可能是由宿主、环境和病原体因素共同作用的结果,这些因素的相互作用导致了流行病学特征、病原体适应性或耐受性发生变化。

六、诺卡氏菌

(一)病原学

1888年Edmond Nocardia从一例牛皮疽病例中首次分离出一种需氧的丝状微生物;1889年Trevisan将此种微生物命名为鼻疽诺卡氏菌(*Nocardia farcinica*),并创立了诺卡氏菌属。1890年Eppinger又从人的致死性大脑脓肿中发现类似的带分枝的丝状微生物,命名为星状分枝丝菌（*Cladothrix asteroides*）；1896年将其重新命名为星状诺卡氏菌（*Nocardia asteroides*）。随后，人们从患有严重感染的人和动物身上分离出越来越多的诺卡氏菌种类。诺卡氏菌属于细菌界，放线菌门，放线菌纲，放线菌亚纲，放线菌目，棒状杆菌亚目，诺卡氏菌科，诺卡氏菌属。诺卡氏菌是一种腐生细菌，在自然界中广泛分布，呈世界性分布，可存在于淡水、海水、土壤、灰尘、腐烂物和动物排泄物等之中。

诺卡氏菌的菌体细长，带有分枝，培养时间长时菌丝可断裂成杆状、球菌状或链球菌样的孢子或孢子丝。革兰氏染色呈阳性，由于致病性菌株细胞壁含分枝菌酸，故抗酸染色为阳性。诺卡氏菌为严格需氧菌，能够形成气生菌丝。其营养要求不高，在普通培养基上或真菌培养基上，于室温或37℃均能生长。然而诺卡氏菌的增殖速度较慢，从临床样本中初次分离培养通常需要约1周的培养时间才能见到明显的菌落。正因诺卡氏菌生长缓慢，在未观察到肉眼可见的诺卡氏菌菌落前就将培养基丢弃，致使诺卡氏菌常常不被人们发觉。诺卡氏菌种类繁杂，不同种类的诺卡氏菌在不同的培养基上呈现出不同的菌落形态。不同种类的诺卡氏菌可产生不同的类胡萝卜样色素，从而使菌落颜色各异；也可产生黑色素样的水溶性色素，使菌落周围的培养基呈棕色。一般来说，在固体培养基上，诺卡氏菌菌落干燥、不透明，表面粗糙呈疣状，致密且硬，菌落基部嵌入培养基。在液体培养基中，诺卡氏菌可形成菌膜样结构，浮于液面；同时形成菌体凝块，沉于肉汤底部，而液体保持澄清。

（二）形态学检查

支气管冲洗液、痰、脓、引流液、组织、脑脊液或者乳汁等，均为最常见的临床样本[图1-9中的（a）和（b）]。获取这些样本后，首先查看样本中有无颗粒样物质；若存在颗粒物，需用无菌生理盐水小心冲洗，并将颗粒物碾碎后通过显微镜进行观察。在放线菌和某些诺卡氏菌的感染中常常能够看到颗粒物。倘若高度怀疑存在这类微生物，后续可进行一些具有针对性的检测与细菌培养。样本的宏观和微观检查是得出明确诊断的首要步骤。将样本涂片进行抗酸染色和革兰氏染色，在显微镜下能观察到革兰氏阳性、细长的、有分枝的丝状微生物[图1-9中的（c）和（d）]。抗酸染色中，使用弱酸（1%硫酸）处理后，诺卡氏菌菌丝通常既呈现抗酸染色阳性，又呈现抗酸染色阴性。这种可变的结果或许与培养时所使用的培养基种类、诺卡氏菌培养的时间相关。因此对样本直接进行抗酸染色来判定结果是不可靠的。抗酸染色仅能用于革兰氏染色之后对结果的进一步证实。在培养基上观察到明显的诺卡氏菌菌落通常至少需要培养 28~72 h。用体视镜观察,在诺卡氏菌菌落表面可见"棉花糖"样外观。不同的诺卡氏菌具有不同的菌落形态：*N. farcinica* 菌落由起初的光滑的、细菌样外观变为橙色；*N. brasiliensis* 菌落为典型的黄色；*N. asteroides* type Ⅵ 通常能够产生棕色色素,这在其他种类的诺卡氏菌中是极为罕见的；而其他种类的诺卡氏菌菌落大多呈白垩样。

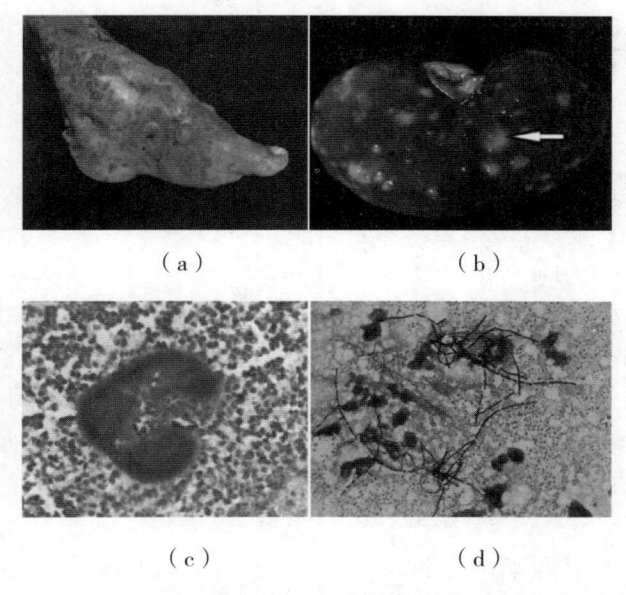

(a) (b)

(c) (d)

图 1-9 诺卡氏菌病的发病特点

(三)生理生化特性

鉴定诺卡氏菌种属的传统方法是开展一系列的生化试验,涵盖乙酰胺、酪蛋白、酪氨酸、黄嘌呤和次黄嘌呤等的利用能力。*N. farcinica* 在 45℃ 培养 72 h 就达到成熟状态,能够迅速(48 h 内)水解乙酰胺作为碳源和氮源,还能够快速分解鼠李糖并产酸,同时不存在芳香基硫酸酯酶活性。*N. nova* 无法分解鼠李糖和乙酰胺,不能在 45℃ 下生长；在培养基上培养两周以后会产生芳香基硫酸酯酶活性。*N. asteroides* 不能在 45℃ 下生长,不能利用乙酰胺,也没有芳香基硫酸酯酶活性,然而 N. asteroids Ⅵ型能够在45℃下生长。N. abscessus 和 N. nova 都能产生硝酸盐还原酶和尿素酶,但都不能在45℃下生长,然而,它们之间也存在差异: N. abscessus 能够利用柠檬酸,而 N. nova 不能。另外,N. cyriacigeorgica 和 N. abscessus 都能水解七叶苷和产生 α-葡糖苷酶,但是 N. cyriacigeorgica 不能产生尿素酶。Deanna 等人借助已有的 API 20C AUX 鉴定系统,结合几种生化试验和纸片扩散法药物敏感性实验,构建了一种新的、精确的诺卡氏菌鉴定方法,成功地将75株诺卡氏菌分离株归类为8个诺卡氏菌种。

七、支原体

（一）病原学

支原体属涵盖100多个种,其中能够感染牛的主要包括牛支原体(*Mycoplasma bovis*)、加利福尼亚支原体、牛生殖支原体、产碱支原体、加拿大支原体等。其中,牛支原体是对牛致病性最强、造成经济损失最大的物种。牛支原体归属于原核生物界、厚壁菌门、柔膜菌纲、支原体目、支原体科、支原体属。支原体是当前发现的最小且结构最简单的原核生物,也是能在无生命培养基中自行繁殖的最小微生物。

（二）支原体型奶牛乳腺炎

1961年,牛支原体被证实是引发美国奶牛乳腺炎病例的致病微生物,并首次被人类分离鉴定为奶牛乳腺炎致病菌。随着牛群的迁徙,牛支原体广泛存在于北美的各大牛场(1997年),此后许多国家相继分离出了牛支原体:以色列(1964年),西班牙(1967年),澳大利亚(1970年),法国(1974年),英国和捷克斯洛伐克(1975年),德国(1977年),丹麦(1981年),瑞士(1983年),墨西哥(1988年),韩国和巴西(1989年),北爱尔兰(1993年),爱尔兰(1994年)和智利(2000年)。截至当下,除挪威等少数国家外,世界范围内的大多数国家都检测到了该病原。在美国、澳大利亚和欧洲,支原体导致的奶牛乳腺炎引发了严重的经济损失。在美国,约超过50%的支原体乳腺炎由牛支原体引起。在美国不同地区牛支原体的流行率各异,东部地区大罐奶中支原体阳性率约为3%,而西部地区大罐奶阳性率为9.4%。在欧洲,牛支原体是最主要的乳腺炎致病性支原体,约1/4的临床型乳腺炎以及40%的亚临床型乳腺炎被认为与支原体相关。有研究表明,在北美和欧洲,大罐奶中支原体的流行率为1%~8%。在导致支原体相关疾病的支原体成员中,牛支原体是工业化国家致病力最强的支原体。在墨西哥,支原体乳腺炎的流行率为55%。在伊朗,牛支原体临床型乳腺炎的流行率为48.75%,而在该研究中48份大罐奶样品中,牛支原体的阳性率为100%。在英国,支原体乳腺炎约占临床型乳腺炎的1%。在法国,牛支原体在大罐奶中的分离率不高于1%。在比利时,1.5%的大罐奶为牛支原体阳性,希腊为5.4%。在我国,由牛支原体引发的奶牛乳腺炎最早于1983年被首次报道。随着我国规模化养牛业的发展,牛支原体已成为全国分布的重要病原之一。2008年牛支原体首次从引起湖北地区犊牛肺炎的病例中被鉴定报道,随后乳腺炎的病例报告也偶有出现。支原体乳腺炎属于传染性疾病,因此针对传染性疾病的防控策略对牛支原体乳腺炎同样有效。支原体乳腺炎的传染源主要包括病牛、共享水源。传播媒介主要涵盖挤奶工手套、挤奶设备和消毒巾等。挤奶前严格系统的卫生管理是防控支原体乳腺炎的有效策略,包括采用单独的服务毛巾挤奶、挤奶者使用的手套消毒、强化挤奶后装置的消毒和挤奶后奶头的药浴。

(三)支原体型奶牛乳腺炎的传播机制及对奶牛的影响

牛支原体乳腺炎传染性极强,单乳区被检测出牛支原体阳性后,在极短时间内即可发展成多乳区感染。奶牛感染牛支原体后,会导致产奶量急剧下降,牛奶的质量下降,奶外观异常,有时伴有脓性分泌物,但是单纯感染牛支原体的乳区,奶样不会如其他常规细菌感染一样产生气味变臭的情况。由于牛支原体感染通常演变成慢性感染,因此牛支原体引起的乳腺炎不像常规细菌性乳腺炎容易被在线兽医及时发现。将牛支原体感染牛引入牛群中是导致牛群间牛支原体传播的主要途径。牛鼻与牛鼻之间直接接触,或通过水池等媒介间接接触,或将灭菌不完全的牛奶饲喂犊牛,以及不良的卫生管理等均是导致牛支原体在牛群内传播的主要原因。牛支原体多易在黏膜类组织器官感染,一旦成功建立感染,则易导致长期持续性感染,并出现持续性或间歇性排菌。犊牛可通过接触成年奶牛或饮用未经完全消毒的牛奶感染牛支原体。牛支原体乳腺炎可由该病原通过乳导管外源途径感染,或者在呼吸道由其他组织繁殖位点经血液细胞携带,移行至奶牛乳腺处造成感染。宿主往往无法有效清除该病原,多数牛经历临床型后,转变为慢性感染或持续性感染,犊牛生长缓慢或消瘦,成年奶牛生产性能下降。传染性病原微生物进化出多种策略干扰宿主免疫系统,发展成慢性感染。然而,人们对牛支原体的感染、传播、病原体在宿主体内存活与增殖,以及免疫逃避的机制尚缺乏深入了解。

(四)支原体型奶牛乳腺炎的影响因素

牛支原体乳腺炎的发生与牛场规模具有显著的相关性,规模大于500头的规模化奶牛场更易发生支原体性乳腺炎。在一项研究中,研究者对664个牛场进行了长达2.25年的风险分析,结果显示,在小规模奶牛场中,大罐奶的体细胞数与牛支原体感染不相关,而在大型牛场中,奶的产量与牛支原体感染显著相关。另外,有研究表明,牛场规模与大罐奶中的牛支原体阳性率显著相关。McCluskey等人通过分析不同规模牛场与大罐奶牛支原体阳性率的关系,发现规模小于100头、100头至499头和大于500头的牛场大罐奶阳性率分别为2.1%、3.9%和21.7%。奶牛场规模一方面影响卫生管理水平,提高了易感宿主的感染概率,同时,多数国家奶牛场不能严格坚持自繁自养,通过引入商品牛扩大规模也是导致奶牛场规模更易出现牛支原体相关疾病的重要因素。牛群的规模越小,支原体的传播途径越容易被阻断。奶牛场引入新牛时,如果新牛作为传染性病原携带者,牛支原体的新发感染总是能追溯到与发病牛接触或牛群引入无症状感染牛事件。无症状感染牛在运输到新的目的地后,由于运输应激或者经历分娩等应激后,发病率相对升高。一旦感染牛进入新的牛群,牛支原体的发病率通常可达40%。同时,感染牛群常伴随着亚临床乳腺炎,也可能以关节炎或肺炎形式发生严重的疫情。

干奶期奶牛发生牛支原体乳腺感染的概率相对较小。但是,这种乳腺炎可持续数月,并可能会发展成泌乳期乳腺炎,进而感染新生牛。在英国,2015年在局部地区不同牛场发生了四次无关联性的干奶期牛支原体乳腺炎事件。主要表现为干奶期10天至5周后发病,多乳区同时感染。虽然进行了正确的干奶期处理，如乳腺内注入组合抗生素并使用封奶剂，同时经过追溯核查，确认卫生管理措施不存在明显的错误。但干奶期奶牛牛支原体乳腺炎的发生仍更接近于散发。另外，其他众多风险因素也给牛支原体的防控带来困难。其中，缺乏医疗隔离区也是牛支原体牛群内传播的重要风险因素。医疗隔离区病牛在治疗痊愈后，应在归群之前做支原体检测，因为牛支原体感染牛可长期向环境中排出病原体。犊牛在产后应及时与成年母牛隔离，以防止通过牛鼻对牛鼻的直接接触以及气溶胶会导致犊牛感染牛支原体。

八、格氏乳球菌

（一）病原学

格氏乳球菌是一种革兰氏阳性、球形的细菌，归属于链球菌科，乳球菌属。格氏乳球菌起初分属于链球菌属，因其在形态、生理和生化特性方面与链球菌属相似，随着分子生物学技术的发展，核酸杂交和超氧化物歧化酶的免疫关系差异，部分链球菌属细菌被重新划至乳球菌属，格氏乳球菌便是其中之一，它通常存在于淡水、海水以及乳制品当中。

（二）格氏乳球菌的毒力基因

格氏乳球菌的致病机理尚未完全明晰，目前已发现的相关因素包括溶血素、LPxTG蛋白、铁载体和各类酶。溶血素基因编码的蛋白能够溶解红细胞，在感染期间可能与分泌的蛋白酶协同作用以促进宿主组织的破坏。LPxTG蛋白是一类广泛存在于细菌表面的蛋白质，其命名源于其保守的氨基酸序列特征（即带有LPxTG结构的亚基），其中L代表Leucine（亮氨酸），P代表Proline（脯氨酸），T代表Threonine（苏氨酸），G代表Glycine（甘氨酸），能够协助细菌在宿主细胞上定位、附着和侵入。分选酶（Sortase）是一种普遍存在于革兰氏阳性细菌中的转肽酶，能够与宿主细胞表面的特定蛋白质相结合。NADH氧化酶和超氧化物歧化酶（SOD）是帮助病原体在有氧环境中存活的酶，细菌通过产生这些酶来保护自身，避免被消灭。

(三)格氏乳球菌型奶牛乳腺炎

1983年,格氏乳球菌首次在奶牛乳腺炎病例中被鉴定出来,随后出现众多报道:西班牙的水牛亚临床乳腺炎,比利时的奶牛亚临床型乳腺炎,美国乳球菌属乳腺炎的暴发。从2017年4月至2021年7月,我们从全国各省份收集的2899份临床型奶牛乳腺炎样品中分离出39株格氏乳球菌,总分离率为1.35%。2017年,从680份奶样中未分离出格氏乳球菌;2018年,从553份奶样中分离出格氏乳球菌5株,分离率0.9%;2019年与2018年情况相同;2020年,从428份奶样中分离出格氏乳球菌17株,分离率为4.0%;截至2021年7月,从685份奶样中分离出格氏乳球菌12株,分离率1.8%。格氏乳球菌的分离率有上升的趋势,其中39株分离株来自全国6个北方省份,河北15株,宁夏12株,山西3株,山东4株,内蒙古3株,甘肃2株。格氏乳球菌引起的乳腺炎通常是亚临床型、环境性的,这意味着感染的奶牛可能没有明显的症状。然而,感染的奶牛可能因体细胞计数增加,存在炎症,从而致使乳品质量和数量下降。在严重的情况下,奶牛可能表现出乳房肿胀、红肿和疼痛等临床症状。研究发现沙床能够成为格氏乳球菌的储存库,格氏乳球菌可以通过沙床传染给奶牛,然后通过感染的乳房进入牛奶。

九、浅绿气球菌

(一)病原学

浅绿气球菌($Aerococcus\ viridans$)是一种革兰氏阳性球菌,隶属于链球菌科气球菌。浅绿气球菌通常形成透明或半透明的圆形菌落,菌落直径2~3 mm,菌落能在绵羊血平板上形成一圈浅绿色的溶血环,一般呈触酶阴性和氧化酶阴性,以上表型特点与链球菌属的细菌极为相似,所以这种细菌常被误认为是链球菌,这或许是这种细菌被低估的原因之一。浅绿气球菌在绵羊血—胰酪大豆胨琼脂平板上呈圆形中等大小(直径为1~3 mm),边缘光滑、表面湿润且菌落周围有α浅绿色溶血,菌落形态和溶血都与链球菌非常类似。经革兰氏染色,镜检呈革兰氏阳性，菌体为球形或椭球形，排列形状较为特殊，呈两个或四个互相连接（图1–10）；扫描电镜结果显示菌体为椭球形，直径为1.5～2 μm。浅绿气球菌一般被认为是一种机会致病菌，广泛存在于环境的灰尘和空气中，在医院和牛场的环境中也有广泛分布。这种细菌最初受到关注主要是因其引起龙虾和鱼类等水产动物的大面积死亡，给水产业造成巨大的经济损失。浅绿气球菌最早在加拿大境内大西洋中的龙虾体内分离得出，能够引发龙虾的"高夫卡败血症"，常在龙虾的运输途中从破损的虾壳入侵至龙虾体内导致龙虾死亡，给龙虾捕捞业带来巨大的经济损失。另外，浅绿气球菌还能引起其他水产动物的致死性疾病，我国广东一个渔场于2010年因浅绿气球菌的感染而损失了30%～40%的罗非鱼。

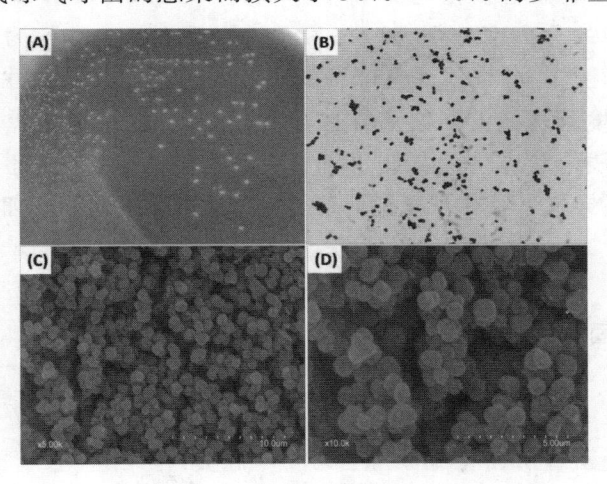

图1–10　浅绿气球菌形态学特征

注：（A）浅绿气球菌于绵羊血–TSA培养基上生长形态为菌落呈白色或浅黄色，圆形，直径为1～3 mm，表面湿润且有α浅绿色溶血环。（B）革兰氏染色为阳性球菌，两两或者四个菌体互相连接。（C）和（D）扫描电镜下菌体呈直径1.5～2 μm的椭球形。

（二）浅绿气球菌型奶牛乳腺炎

浅绿气球菌广泛存在于牛场环境中，在奶牛皮肤上亦有分布，因此尽管这种细菌在大罐奶和亚临床奶牛乳腺炎病例中曾被分离鉴定，但因其致病性不明，常被认为是采样或挤奶时从环境进入牛奶，不会对奶牛乳腺健康构成威胁，故而未引起牛场管理人员和科研工作者的重视。近年来，数个国家均有报道从临床和亚临床乳腺炎奶样中分离到浅绿气球菌,这再次将浅绿气球菌与奶牛乳腺炎关联起来。我们也提出一个假设,浅绿气球菌或许是奶牛乳腺炎病原菌的一种,能够危害奶牛乳腺健康;然而至今,尚无关于浅绿气球菌对奶牛乳腺健康的相关研究报道。另外,据研究结果显示,虾源性浅绿气球菌可分为有毒株和无毒株,且有毒株和无毒株在一定条件下能够相互转化。从 19 个大型牧场共采集 1008 份亚临床乳腺炎奶样,其中 10 个牧场中共分离鉴定 60 株浅绿气球菌,总分离率为 5.9%;河北的一个牧场分离率最高,从 34 份亚临床乳腺炎奶样中分离到 10 株菌,分离率为 29.4%;北京和天津各牧场的分离率差异较大,北京三个牧场的分离率分别为 5.4%、23.0% 和 23.1%,而天津三个牧场的分离率分别为 5.8%、8.3% 和 20.5%;河北、北京和天津的分离率分别为 15.5%、15.0% 和 11.4%。作为一种环境菌和机会致病菌,牛源性浅绿气球菌的菌株间可能存在致病性和毒力的差异,由于环境菌株间的生物多样性较大,而菌株间的毒力差异也可能较大。另外,浅绿气球菌除与奶牛乳腺炎联系较为紧密外,还有研究报道这种病原菌导致了犊牛严重的呼吸道感染,并且强调了该细菌的多重耐药性。

十、凝固酶阴性葡萄球菌

(一)凝固酶阴性葡萄球菌型奶牛乳腺炎

凝固酶阴性葡萄球菌作为环境性病原菌,多年来一直被人忽视。然而,近些年来因其检出率高且会引起产奶量下降而备受关注,该类别病原体在众多国家已成为奶牛乳腺炎的主要致病菌。凝固酶阴性葡萄球菌在牛场分布广泛,如牛体表与环境中皆有存在,尽管这类微生物在乳腺内仅能引发轻微炎症,却能使产奶量降低超过 8%。在已鉴定出的相关凝固酶阴性葡萄球菌的病原菌约有 20 多种。其中报道过与奶牛乳腺炎相关的有表皮葡萄球菌、腐生葡萄球菌、溶血葡萄球菌、木糖葡萄球菌、猪葡萄球菌、模仿葡萄球菌等。当动物机体免疫力低下时,凝固酶阴性葡萄球菌进入乳房可引发感染。目前对于凝固酶阴性葡萄球菌的观点存在分歧。凝固酶阴性葡萄球菌曾一度被认为只会导致隐性乳腺炎和温和型临床乳腺炎,虽然有时也会造成严重症状,但相比金黄色葡萄球菌等传统的主要病原菌重要性较弱。然而，经产牛通常在哺乳后期感染凝固酶阴性葡萄球菌，初次分娩奶牛通常在产犊前或者产犊后短时间内感染。凝固酶阴性葡萄球菌主要致使产犊前后隐性乳腺炎的发生，哺乳期的凝固酶阴性葡萄球菌感染通常伴随体细胞数增加，同时可造成持续性感染，导致牛奶品质降低。甚至有研究指出，凝固酶阴性葡萄球菌感染可能致使母牛流产。

在德国，35%的隐性乳腺炎乳区被检测出凝固酶阴性葡萄球菌；在美国田纳西州，高体细胞数奶牛群中凝固酶阴性葡萄球菌的平均感染率为28%；而早前的美国和加拿大有15%的产后牛乳房感染均为凝固酶阴性葡萄球菌。在爱沙尼亚的一项调查中，乳区的凝固酶阴性葡萄球菌阳性率为16%；而在芬兰，这一比例更高达50%。不过不同国家的结果难以相互对比，因为每个国家用于细菌实验的标准菌落形成单位不尽相同。实际上，隐性乳腺炎的凝固酶阴性葡萄球菌分离率普遍高于临床乳腺炎。凝固酶阴性葡萄球菌包括47个不同的种系以及24个亚种，其中在牛奶中可分离出的大概有十几种。在奶牛乳腺炎的病因方面，凝固酶阴性葡萄球菌的流行病学及其中种类的关联性仍存争议。

（二）凝固酶阴性葡萄球菌的分类

凝固酶阴性葡萄球菌种类繁多，不同饲养环境、管理的牧场凝固酶阴性葡萄球菌的种类差异较大。De Visscher等人的调查显示，产色葡萄球菌、松鼠葡萄球菌和科氏葡萄球菌是分离率最高的种类；Trembla等人在加拿大的研究表明，产色葡萄球菌、模仿葡萄球菌、木糖葡萄球菌、溶血葡萄球菌、表皮葡萄球菌是主要的优势菌株，这5个种类代表了分离菌株的85%。另外产犊前乳头顶端不洁净更易引起科氏葡萄球菌、马胃葡萄球菌、腐生葡萄球菌或者松鼠葡萄球菌的感染，这些病原菌主要源自自然环境。相比未受感染的乳区，产色葡萄球菌、模仿葡萄球菌和木糖葡萄球菌与体细胞计数增高关系密切。Satu等人同样发现最常被分离到的凝固酶阴性葡萄球菌是产色葡萄球菌和模仿葡萄球菌，溶血葡萄球菌和表皮葡萄球菌分离率也较高。

1. 表皮葡萄球菌

表皮葡萄球菌(*Staphylococcus epidermidis*)是滋生于生物体表皮的一种革兰氏阳性球菌,存在于人体的皮肤、阴道等部位,因常堆聚成葡萄串状,故而得名。表皮葡萄球菌一般不被噬菌体裂解,依据生化反应和产生酶类的不同可分为 6 型。表皮葡萄球菌正常存在于人的皮肤、鼻腔及肠道中。不产生血浆凝固酶,也不产生 α 溶血素等毒性物质,故通常认为是非致病性葡萄球菌。但近年来该菌的致病性逐渐受到重视。表皮葡萄球菌直径为 0.5 ~ 1.5 μm,其营养要求不高,在普通培养基上生长良好。需氧或兼性厌氧,最适生长温度为 37℃,最适 pH 为 7.4。具有耐盐性,生长时需要生物素。在肉汤培养基中经 37℃、24 h 孵育后,呈均匀混浊生长。在普通琼脂平板上形成凸起、光滑不透明的圆形菌落,直径为 1 ~ 2 mm。表皮葡萄球菌一般产生白色或柠檬色色素,故菌落呈现白色或柠檬色。自 1955 年起,在北美陆续发现少数产生紫色色素的表皮葡萄球菌。在有氧或无氧条件下,表皮葡萄球菌都可分解葡萄糖,产酸不产气。一般也分解乳糖和麦芽糖产酸。在厌氧条件下,一般不分解甘露醇。能还原硝酸盐。可以通过精氨酸水解酶分解精氨酸产氨。水解马尿酸,不能水解溶素或鸟氨酸。很多表皮葡萄球菌株都产生磷酸酯、脂蛋白类脂酶和酯酶。有些菌株可产生溶菌酶和不耐热的核酸酶。表皮葡萄球菌包括白色葡萄球菌和柠檬色葡萄球菌。在普通琼脂平板培养基上生长的菌落产生白色脂色素,呈白色菌落者称为白色葡萄球菌,而产生柠檬色色素,使菌落呈柠檬色者称为柠檬色葡萄球菌。

葡萄球菌抗原构造复杂,已发现的在 30 种以上,了解其化学组成及生物学活性的仅少数几种。葡萄球菌 A 蛋白(SPA)是存在于菌细胞壁的一种表面蛋白,位于菌体表面,与胞壁的粘肽相结合。它与人及多种哺乳动物血清中的 IgG 的 Fc 段结合,因而可用含 SPA 的葡萄球菌作为载体,结合特异性抗体进行协同凝集试验。A 蛋白有抗吞噬作用,还有激活补体替代途径等活性。SPA 是一种单链多肽,与细胞壁肽聚糖呈共价结合,是完全抗原,具有属特异性。多糖抗原具有群特异性,存在于细胞壁,借此可以分群,A 群多糖抗原体化学组成为磷壁酸中的 $N-$ 乙酰葡胺核糖醇残基。B 群化学组成是磷壁酸中的 $N-$ 乙酰区糖胺甘油残基。几乎所有金黄色葡萄球菌菌株的表面有荚膜多糖抗原的存在,而表皮葡萄球菌仅个别菌株有此抗原。

表皮葡萄球菌很重要的毒力因子就是生物膜,对表皮葡萄球菌生物膜的研究尤为重要。表皮葡萄球菌生物膜的形成已被普遍认为是导致奶牛乳腺炎的重要毒性机制,促进表皮葡萄球菌在乳腺内的长期感染。形成生物膜是细菌的一种特性,并且与基因相关,代表了乳腺炎发病机制的一种选择性优势。生物膜的形成是细菌在自然环境中生存的一种重要模式。一旦形成这种生存模式,可能会失去对各种抗生素的敏感性,对机体调节吞噬机制以及传统抗生素药物的耐受力会大大增强,相比未形成生物膜的相同种类其他菌株,对抗生素的敏感性会降低100～1000倍。微生物这种改变的耐药性是慢性感染的一个重要因素。同时,形成生物膜是一个动态的过程,表现为脱落的细胞可能即时、急速增殖,然后占据其他的表面。这个脱落的过程会导致基因表达的变化、毒力因子和黏附因子改变,以及快速增殖等,这些通常与感染的临床症状的再出现有关。生物膜的形成也能对宿主机体组织产生伤害,因为生物膜可以刺激白细胞释放溶酶体酶类、活性氧以及各种氮类。总之,生物膜代表了微生物在乳腺内得以生存的一种机制,使得牧场的病原菌得以长期存在。因此,研究生物膜以及其他与主要病原菌增殖并持续感染的能力有关的毒力因子至关重要。

表皮葡萄球菌早在20世纪80年代在临床乳腺炎分离菌株中的比例就有1.7%。目前,表皮葡萄球菌是亚临床乳腺炎中较常被分离到的凝固酶阴性葡萄球菌。相比其他与奶牛乳腺炎有关的凝固酶阴性葡萄球菌种类,表皮葡萄球菌并非奶牛皮肤或者黏膜上的常驻菌群,然而却是人类皮肤上较为普遍的一种细菌,有研究者认为其是人类的专性寄生菌,因此可以猜测导致奶牛乳腺炎感染的表皮葡萄球菌实际上可能来自人类。

2. 松鼠葡萄球菌

松鼠葡萄球菌(*Staphylococcus sciuri*)是一种革兰氏阳性菌,属于凝固酶阴性葡萄球菌,直径0.8～1.0 μm,镜下观察呈葡萄状排列,有些菌株还具有荚膜或黏层。在普通营养琼脂培养基上形成直径1～2 mm、表面光滑、边缘整齐、圆形凸起的白色或乳白色菌落;在血琼脂平板上形成圆形或卵圆形、不溶血、白色或淡黄色的菌落。松鼠葡萄球菌首次由

大肠杆菌感染致奶牛乳腺炎的致病机制及防治

Kools 等人于 1976 年报道，从人类及动物皮肤中分离得到；此后，松鼠葡萄球菌不断从各个宿主分离，包括有袋动物、啮齿动物、食肉动物、猴子、鲸目动物和家畜在内的多种野生动物及家养动物，其中最重要的宿主包括牛、羊、马和狗等宠物与经济动物；此外，由于松鼠葡萄球菌可以以无机氮盐作为唯一的氮源生长，目前已能从土壤、沙子、水和沼泽草中分离出来，可见其广泛的宿主及分布范围。与金黄色葡萄球菌相比，松鼠葡萄球菌在宿主正常的情况下，能在皮肤表面生存，其致病性较低，此前常被人们忽视。但近些年来，在畜牧养殖与人类医学中，松鼠葡萄球菌逐渐受到各国学者的重视，被认为是机会致病菌，在特定情况下，也会引起多种类型的感染。目前报道了人心内膜炎感染、腹膜炎、感染性休克、尿路感染、眼内炎、盆腔炎等病例；在兽医领域，有仔猪渗出性皮炎、肺部病变和心内膜炎，以及牛乳腺炎感染等病例。此外，由于其广泛的宿主适应性，人畜之间相互传播的风险也大大高于一些专性致病菌。

松鼠葡萄球菌会产生 DNA 酶、明胶酶、脂肪酶、磷酸酶和蛋白酶等，作用于宿主细胞产生致病性。此外，松鼠葡萄球菌毒力因子还包括编码亮白细胞素（lukS）、纤连蛋白结合蛋白 A（fnbA）、免疫逃避因子（isdA）和骨唾液蛋白结合蛋白（bbp）基因，其质粒上还存在多种能引起宿主应激反应的基因，包括渗透、氧化、热/冷休克和周质应激相关基因等。其携带的表皮角质剥脱毒素基因（Exh-C）是导致仔猪渗出性内膜炎的关键毒力因子，会编码产生 rExh-C 二聚体，这是一种调节巨噬细胞功能的物质，能抑制巨噬细胞的吞噬作用、刺激巨噬细胞产生 IL-6、IL-12、TNF-α 和 NO，导致实验小鼠皮肤脱落，在仔猪上引起严重致死性的渗出性表皮炎。

3. 产色葡萄球菌

产色葡萄球菌是 Staphylococcus 属的微生物，为革兰氏阳性杆菌，呈球杆状，单个细胞大小为 1.0 ~ 1.5 μm × 1.5 ~ 2.5 μm；专性需氧；氧化酶阴性，无动力；硝酸盐试验阴性；最适生长温度为 30 ~ 32℃。产色葡萄球菌广泛分布于自然界和营养环境，是临床标本中分离的第二个最常见的非发酵菌，还能存在于健康人的皮肤表面。

参考文献

[1] BOGNI C, ODIERNO L, RASPANTI C, et al.War against mastitis: Current concepts on controlling bovine mastitis pathogens[J].Science against Microbial Pathogens: Communicating Current Research and Technological Advances,2011: 483-494.

[2] BURVENICH C, VAN MERRIS V, Mehrzad J, et al.Severity of E.Coli mastitis is mainly determined by cow factors[J].Veterinary Research,2003,34（5）: 521-564.

[3] HOGEVEEN H, HUIJPS K, LAM T J G M. Economic aspects of mastitis: New developments[J].New Zealand Veterinary Journal,2011,59（1）: 16-23.

[4] ALI T, RAHMAN S U, ZHANG L M, et al.ESBL-producing Escherichia coli from cows suffering mastitis in china contain clinical class 1 integrons with CTX-M linked to ISCR1[J].Frontiers in Microbiology,2016,7: 1931.

[5] KEMPF F, SLUGOCKI C, BLUM S E, et al.Genomic comparative study of bovine mastitis escherichia coli[J].PLoS One,2016,11（1）: e0147954.

[6] GAO J, BARKEMA H W, ZHANG L M, et al.Incidence of clinical mastitis and distribution of pathogens on large Chinese dairy farms[J].Journal of Dairy Science,2017,100（6）: 4797-4806.

[7] Escherich T.Dıё Darmbacterien des Neugeborenen und Saüglings[J].Fortschritte Der Medizin,1885,3（16）: 515-522.

[8] HACKER J, BLUM-OEHLER G, HOCHHUT B, et al.The molecular basis of infectious diseases: Pathogenicity Islands and other mobile genetic elements. A review[J].Acta Microbiologica et Immunologica Hungarica,2003,50（4）: 321-330.

[9] HALASA T, HUIJPS, K, OSTERAS, O, et al.Economic effects of bovine mastitis and mastitis management: A review[J].The Veterinary Quarterly,2007,29（1）: 18-31.

[10] TENAILLON O, SKURNIK D, PICARD B, et al.The population

genetics of commensal Escherichia coli[J].Nature Reviews Microbiology,2010,8（3）:207-217.

[11] CHAUDHURI R R, HENDERSON I R.The evolution of the Escherichia coli phylogeny[J].Infection, Genetics and Evolution, 2012,12（2）:214-226.

[12] TOUCHON M, HOEDE C, TENAILLON O, et al.Organised genome dynamics in the Escherichia coli species results in highly diverse adaptive paths[J].PLoS Genetics,2009,5（1）:e1000344.

[13] PETZL W, ZERBE H, GÜNTHER J, et al.Escherichia coli, but not Staphylococcus aureus triggers an early increased expression of factors contributing to the innate immune defense in the udder of the cow[J].Veternary Research,2008,39（2）:18.

[14] XU G F, AN W, WANG H D, et al.Prevalence and characteristics of extended-spectrum β-lactamase genes in Escherichia coli isolated from piglets with post-weaning diarrhea in Heilongjiang Province, China[J].Frontiers in Microbiology,2015,6:1103.

[15] LIU Y X, LIU G, LIU W J, et al.Phylogenetic group, virulence factors and antimicrobial resistance of Escherichia coli associated with bovine mastitis[J].Research in Microbiology,2014,165（4）:273-277.

[16] GHANBARPOUR R, OSWALD E.Phylogenetic distribution of virulence genes in Escherichia coli isolated from bovine mastitis in Iran[J].Research in Veterinary Science,2010,88（1）:6-10.

[17] NATARO J P, KAPER J B.Diarrheagenic Escherichia coli[J].Clinical Microbiology Reviews,1998,11（1）:142-201.

[18] DYER J G, SRIRANGANATHAN N, NICKERSON S C, et al.Curli production and genetic relationships among Escherichia coli from cases of bovine mastitis[J].Journal of Dairy Science,2007,90（1）:193-201.

[19] SRINIVASAN V, GILLESPIE B E, LEWIS M J, et al.Phenotypic and genotypic antimicrobial resistance patterns of Escherichia coli isolated from dairy cows with mastitis[J].Veterinary Microbiology, 2007,124(3/4):319-328.

[20] KEMPF F, SLUGOCKI C, BLUM S E, et al.Genomic comparative study of bovine mastitis Escherichia coli[J].PLoS One,2016,11(1): e0147954.

[21] SHPIGEL N Y, LEVIN D, WINKLER M, et al.Efficacy of cefquinome for treatment of cows with mastitis experimentally induced using Escherichia coli[J].Journal of Dairy Science,1997,80 (2): 318-323.

[22] DOGAN B, KLAESSIG S, RISHNIW M, et al.Adherent and invasive Escherichia coli are associated with persistent bovine mastitis[J]. Veterinary Microbiology,2006,116(4): 270-282.

[23] BURVENICH C, VAN MERRIS V, MEHRZAD J, et al.Severity of E.coli mastitis is mainly determined by cow factors[J].Veterinary Research,2003,34(5): 521-564.

[24] SHPIGEL N Y, ELAZAR S, ROSENSHINE I.Mammary pathogenic Escherichia coli[J].Current Opinion in Microbiology,2008,11(1): 60-65.

[25] BRADLEY A J, GREEN M J.A study of the incidence and significance of intramammary enterobacterial infections acquired during the dry period[J].Journal of Dairy Science,2000,83(9): 1957-1965.

[26] SCHUKKEN Y H, BENNETT G J, ZURAKOWSKI M J, et al.Randomized clinical trial to evaluate the efficacy of a 5-day ceftiofur hydrochloride intramammary treatment on nonsevere gram-negative clinical mastitis[J].Journal of Dairy Science,2011,94(12): 6203-6215.

[27] BURVENICH C, VAN MERRIS V, MEHRZAD J, et al.Severity of E.coli mastitis is mainly determined by cow factors[J].Veterinary Research,2003,34(5): 521-564.

[28] SURIYASATHAPORN W, HEUER C, NOORDHUIZEN-STASSEN E N, et al.Hyperketonemia and the impairment of udder defense: A review[J].Veterinary Research,2000,31(4): 397-412.

[29] SMITH K L, HOGAN J S, WEISS W P.Dietary vitamin E and selenium affect mastitis and milk quality[J].Journal of Animal

Science, 1997, 75（6）: 1659-1665.

[30] BRADLEY A J, BREEN J E, PAYNE B, et al.An investigation of the efficacy of a polyvalent mastitis vaccine using different vaccination regimens under field conditions in the United Kingdom[J].Journal of Dairy Science, 2015, 98（3）: 1706-1720.

[31] DOPFER D, BARKEMA H W, LAM T J, et al.Recurrent clinical mastitis caused by Escherichia coli in dairy cows[J].Journal of Dairy Science, 1999, 82（1）: 80-85.

[32] SORDILLO L M, STREICHER K L.Mammary gland immunity and mastitis susceptibility[J].Journal of Mammary Gland Biology and Neoplasia, 2002, 7（2）: 135-146.

[33] WELLNITZ O, BRUCKMAIER R M.The innate immune response of the bovine mammary gland to bacterial infection[J].Veterinary Journal, 2012, 192（2）: 148-152.

[34] ZHAO S, WHITE D G, MCDERMOTT P F, et al. Identifica-tion and expression of cephamycinase bla（CMY）genes in Escherichia coli and Salmonella isolates from food animals and ground meat[J]. Antimicrobial Agents and Chemotherapy, 2001, 45（12）: 3647-3650.

[35] VIGUIER C, ARORA S, GILMARTIN N, et al.Mastitis detection: Current trends and future perspectives[J].Trends in Biotechnology, 2009, 27（8）: 486-493.

[36] LONG E, CAPUCO A V, WOOD D L, et al.Escherichia coli induces apoptosis and proliferation of mammary cells[J].Cell Death and Differentiation, 2001, 8（8）: 808-816.

[37] SHPIGEL N Y, ELAZAR S, ROSENSHINE I.Mammary pathogenic escherichia coli[J].Current Opinion in Microbiology, 2008, 11（1）: 60-65.

[38] 蔡双启,黄莹莹,陈一强.金黄色葡萄球菌毒力因子的研究进展[J].国际呼吸杂志,2015,35（16）: 1242-1245.

[39] KOZIEL J, CHMIEST D, BRYZEK D, et al.The Janus face of alpha-toxin: A potent mediator of cytoprotection in staphylococci-infected macrophages[J].Journal of Innate Immunology, 2015, 7（2）: 187-198.

[40] 陈万义,游春萍,刘振民.金黄色葡萄球菌肠毒素的研究进展[J].乳业科学与技术,2015,38(6):31-37.

[41] GOGOI-TIWARI J, WILLIAMS V, WARYAH C B, et al.Mammary gland pathology subsequent to acute infection with strong versus weak biofilm forming staphylococcus aureus bovine mastitis isolates: A pilot study using non-invasive mouse mastitis model[J]. PLoS One,2017,12(1):e0170668.

[42] 刘庆中,韩立中,孙景勇,等.高水平莫匹罗星耐药MRSA体外生物膜形成能力及相关基因检测[J].中国感染与化疗杂志,2011,11(4):276-280.

[43] 史华英,李青栋,万献尧.金黄色葡萄球菌耐药性与生物膜的相关性研究[J].中华内科杂志,2015,54(12):1063-1065.

[44] TENHAGEN B A, KÖSTER G, WALLMANN J, et al.Prevalence of mastitis pathogens and their resistance against antimicrobial agents in dairy cows in Brandenburg, Germany[J].Journal of Dairy Science,2006,89(7):2542-2551.

[45] MYLLYS V, HONKANEN-BUZALSKI T, VIRTANEN H, et al.Effect of abrasion of teat orifice epithelium on development of bovine staphylococcal mastitis[J].Journal of Dairy Science,1994,77(2):446-452.

[46] GIANNEECHINI R, CONCHA C, RIVERO R, et al.Occurrence of clinical and sub-clinical mastitis in dairy herds in the west littoral region in Uruguay[J].Acta Veterinaria Scandinavica,2002,43(4):221.

[47] ORIEKERINK R G M, BARKEMA H W, VEENSTRA S, et al. Prevalence of contagious mastitis pathogens in bulk tank milk in Prince Edward Island[J].The Canadian Veterinary Journal,2006,47(6):567-572.

[48] BARKEMA, H W, GREEN, M J, BRADLEY, A J, et al. Invited review: The role of contagious disease in udder health[J].Journal of Dairy Science, 2009,92(10):4717-4729.

[49] HIRAMATSU K, CUI L, KURODA M, et al.The emergence and evolution of methicillin-resistant Staphylococcus aureus[J]. Trends

in Microbiology,2001,9（10）: 486-493.

[50] KENAR B, KUYUCUOGLU Y, SEKER E.Antibiotic susceptibility of coagulase-negative staphylococci isolated from bovine subclinical mastitis in Turkey[J].Pakistan Veterinary Journal,2012,32: 390-393.

[51] MATOS J S, WHITE D G, HARMON R J, et al . Isolation of Staphylococcus aureus from sites other than the lactating mammary gland[J].Journal of Dairy Science,1991,74（5）: 1544-1549.

[52] ROBERSON J R, FOX L K, HANCOCK D D, Gay, et al.Ecology of Staphylococcus aureus isolated from various sites on dairy farms[J]. Journal of Dairy Science,1994,77（11）: 3354-3364.

[53] GONZALEZ R N, JASPER D E, FARVER T B, et al.Prevalence of udder infections and mastitis in 50 California dairy herds [J].Journal of the American Veterinary Medical Association,1988,193（3）: 323-328.

[54] WILSON C D, RICHARDS M S.A survey of mastitis in the British dairy herd [J].The Veterinary Record,1980,106（21）: 431-435.

[55] AARESTRUP F M, WEGENER H C, ROSDAHL V T.A comparative study of Staphylococcus aureus strains isolated from bovine subclinical mastitis during 1952-1956 and 1992 [J].Acta Veterinaria Scandinavica,1995,36（2）: 237-243.

[56] SWINKEL S J M, HOGEVEEN H, ZADOKS R N.A partial budget model to estimate economic benefits of lactational treatment of subclinical Staphylococcus aureus mast it is [J]. Journal of Dairy Science,2005,88（12）: 4273-4287.

[57] Ø STERÅS O, HOGEVEEN H, SINGH D, et al . Economic consequences of mastitis[J].Bulletin International Dairy Federation, 2005,394: 2-25.

[58] REKSEN O, SOLVEROD L, BRANSCUM A J, et al.Relationships between milk culture results and treatment for clinical mastitis or culling in Norwegian dairy cattle [J].Journal of Dairy Science,2006, 89（8）: 2928-2937.

[59] BARKEMA H W, SCHUKKEN Y H, ZADOKS R N.Invited

Review：The role of cow, pathogen, and treatment regimen in the therapeutic success of bovine Staphylococcus aureus mastitis[J]. Journal of Dairy Science, 2006, 89（6）: 1877-1895.

[60] GAO J, FERRERI M, YU FQ, et al.Molecular types and antibiotic resistance of Staphylococcus aureus isolates from bovine mastitis in a single herd in China[J].The Veterinary Journal, 2012, 192（3）: 550-552.

[61] DAVIES J.Inactivation of antibiotics and the dissemination of resistance genes[J]. Science, 1994, 264（5157）: 375-382.

[62] LEONARD F C, MARKEY B K. Meticillin-resistant Staphylococcus aureus in animals: a review[J].The Veterinary Journal, 2008, 175（1）: 27-36.

[63] OTTER J A, FRENCH G L. Molecular epidemiology of community-associated meticillin-resistant Staphylococcus aureus in Europe[J]. The Lancet Infectious Diseases, 2010, 10（4）: 227-239.

[64] LEE J H.Methicillin（oxacillin）-resistant Staphylococcus aureus strains isolated from major food animals and their potential transmission to humans[J].Applied and Environmental Microbiology, 2003, 69（11）: 6489-6494.

[65] GARCIA-ALVAREZ L, HOLDEN M T, LINDSAY H, et al.Meticillin- resistant Staphylococcus aureus with a novel mecA homologue in human and bovine populations in the UK and Denmark: A descriptive study[J].The Lancet Infectious Diseases, 2011, 11（8）: 595-603.

[66] VERKADE E, KLUYTMANS J.Livestock-associated Staphylococcus aureus CC398: Animal Reservoirs and human infections[J].Infection, Genetics and Evolution, 2014, 21: 523-530.

[67] 陆承平.兽医微生物学[M].4版.北京: 中国农业出版社, 2007.

[68] 刘海林.猪源肺炎克雷伯氏菌分离菌株耐药性调查及防控技术研究[D].南京: 南京农业大学, 2015.

[69] SIU L K, YEH K M, LIN J C, et al.Klebsiella pneumoniae liver abscess: A new invasive syndrome[J].The Lancet Infectious Diseases, 2012, 12（11）: 881-887.

[70] 赵兴绪,何渊. 奶牛乳腺炎防治 [M]. 北京: 金盾出版社,2007.

[71] LI B, ZHAO Y L, LIU C T, et al.Molecular pathogenesis of klebsiella pneumoniae[J].Future Microbiology,2014,9（9）: 1071-1081.

[72] MAI J Y, ZHU M L, CHEN C, et al.Clinical characteristics of neonatal Klebsiella pneumoniae sepsis and the antibiotic sensitivity pattern of strains[J].Zhongguo Dang Dai Er Ke Za Zhi,2010,12(9): 700-703.

[73] 刘燕,赵京,张成林,等. 肺炎克雷伯氏菌致白颊长臂猿脓肿一例 [J]. 野生动物学报,2011,32（3）: 156-157.

[74] 冯娜. 一株牛源肺炎克雷伯菌的分离鉴定及其基因组初步分析 [D]. 兰州: 甘肃农业大学,2017.

[75] RASMUSSEN B A, BUSH K, KEENEY D, et al.Characterization of IMI-1 beta-lactamase, a class A carbapenem-hydrolyzing enzyme from Enterobacter cloacae[J].Antimicrobial Agents and Chemotherapy,1996,40（9）: 2080-2086.

[76] STRUVE C, BOJER M, NIELSEN E M, et al.Investigation of the putative virulence gene magA in a worldwide collection of 495 Klebsiella isolates: MagA is restricted to the gene cluster of Klebsiella pneumoniae capsule serotype K1[J].Journal of Medical Microbiology, 2005,54（Pt 11）: 1111-1113.

[77] CHUANG Y P, FANG C T, LAI S Y, et al.Genetic determinants of capsular serotype K1 of klebsiella pneumoniae causing primary pyogenic liver abscess[J].The Journal of Infectious Diseases,2006, 193（5）: 645-654.

[78] TURTON J F, PERRY C, ELGOHARI S, et al.PCR characterization and typing of klebsiella pneumoniae using capsular type-specific, variable number tandem repeat and virulence gene targets[J].Journal of Medical Microbiology,2010,59（Pt 5）: 541-547.

[79] ARES M A, FERNÁNDEZ-VÁZQUEZ J L, ROSALES-REYES R, et al.H-NS nucleoid protein controls virulence features of Klebsiella pneumoniae by regulating the expression of type 3 pili and the capsule polysaccharide[J].Frontiers in Cellular and Infection Microbiology,2016,6: 13.

[80] STRUVE C, BOJER M, KROGFELT K A.Identification of a conserved chromosomal region encoding Klebsiella pneumoniae type 1 and type 3 fimbriae and assessment of the role of fimbriae in pathogenicity[J]. Infection and Immunity,2009,77（11）:5016-5024.

[81] DOMENICO P, SALO R J, CROSS A S, et al. Polysaccharide capsule-mediated resistance to ops onophagocytosis in Klebsiella pneumoniae[J].Infection and Immunity,1994,62（10）:4495-4499.

[82] WARD C G, HAMMOND J S, BULLEN J J.Effect of iron compounds on antibacterial function of human polymorphs and plasma[J]. Infection and Immunity,1986,51（3）:723-730.

[83] YU V L, HANSEN D S, KO W C, et al.Virulence characteristics of Klebsiella and clinical manifestations of K.pneumoniae bloodstream infections[J].Emerging Infectious Diseases,2007,13（7）:986-993.

[84] BRISSE S, FEVRE C, PASSET V, et al. Virulent clones of, Klebsiella pneumoniae: Identification and evolutionary scenario based on genomic and phenotypic characterization[J].PLoS One, 2009,4（3）: e4982.

[85] BANNERMAN D D, PAAPE M J, HARE W R, et al. Characterization of the bovine innate immune response to intramammary infection with Klebsiella pneumoniae[J].Journal of Dairy Science,2004,87（8）: 2420-2432.

[86] MUNOZ M A, WELCOME F L, SCHUKKEN Y H, et al.Molecular epidemiology of two Klebsiella pneumoniae mastitis outbreaks on a dairy farm in New York State[J].Journal of Clinical Microbiology, 2007,45（12）: 3964-3971.

[87] GAO J, BARKEMA H W, ZHANG L M, et al.Incidence of clinical mastitis and distribution of pathogens on large Chinese dairy farms[J]. Journal of Dairy Science,2017,100（6）: 4797-4806.

[88] GRÖHN Y T, WILSON D J, GONZÁLEZ R N, et al. Effect of pathogen-specific clinical mastitis on milk yield in dairy cows[J]. Journal of Dairy Science,2004,87（10）: 3358-3374.

[89] NAGASAWA Y, KIKU Y, SUGAWARA K, et al.The bacterial load in

milk is associated with clinical severity in cases of bovine coliform mastitis[J]. The Journal of Veterinary Medical Science, 2019, 81(1): 107-112.

[90] OLDE RIEKERINK R G M, BARKEMA H W, STRYHN H. The effect of season on somatic cell count and the incidence of clinical mastitis[J]. Journal of Dairy Science, 2007, 90(4): 1704-1715.

[91] GAO J, BARKEMA H W, ZHANG L M, et al. Incidence of clinical mastitis and distribution of pathogens on large Chinese dairy farms[J]. Journal of Dairy Science, 2017, 100(6): 4797-4806.

[92] VIGUIER C, ARORA S, GILMARTIN N, et al. Mastitis detection: Current trends and future perspectives[J]. Trends in Biotechnology, 2009, 27(8): 486-493.

[93] HODGES R T, HOLLAND J T, NEILSONT F J, et al. Prototheca zopfii mastitis in a herd of dairy cows[J]. New Zealand Veterinary Journal, 1985, 33(7): 108-111.

[94] BUENO V F F, DE MESQUITA A J, NEVES R B S, et al. Epidemiological and clinical aspects of the first outbreak of bovine mastitis caused by Prototheca zopfii in goias state, Brazil[J]. Mycopathologia, 2006, 161(3): 141-145.

[95] GAO J, ZHANG H Q, HE J Z, et al. Characterization of prototheca zopfii associated with outbreak of bovine clinical mastitis in herd of Beijing, China[J]. Mycopathologia, 2012, 173(4): 275-281.

[96] CORBELLINI L G, DRIEMEIER D, CRUZ C, et al. Ferreiro. Bovine Mastitis due to Prototheca zopfii: Clinical, Epidemiological and Pathological Aspects in a Brazilian Dairy Herd[J]. Tropical Animal Health and Production, 2001, 33(6): 463-470.

[97] COSTA E O, MELVILLE P A, RIBEIRO A R, et al. Epidemiologic study of environmental sources in a Prototheca zopfii outbreak of bovine mastitis[J]. Mycopathologia, 1997, 137(1): 33-36.

[98] DA COSTA E O, RIBEIRO M, RIBEIRO A R, et al. Diagnosis of clinical bovine mastitis by fine needle aspiration followed by staining and scanning electron microscopy in a prototheca zopfii outbreak[J]. Mycopathologia, 2004, 158(1): 81-85.

[99] JAGIELSKI T, BUZZINI P, LASSA H, et al.Multicentre Etest evaluation of in vitro activity of conventional antifungal drugs against European bovine mastitis Prototheca spp. isolates[J].Journal of Antimicrobial Chemotherapy,2012,67（8）: 1945-1947.

[100] JANOSI S, RATZ F, SZIGETI G, et al . Review of the microbiological, pathological, and clinical aspects of bovine mastitis caused by the alga Prototheca zopfii[J].The Veterinary Quarterly, 2001, 23（2）: 58-61.

[101] LASS-FLORL C, MAYR A. Human protothecosis[J]. Clinical Microbiology Reviews,2007,20（2）: 230-242.

[102] MALINOWSKI E, LASSA H, KLOSSOWSKA A . Isolation of Prototheca zopfii from inflamed secretion of udders[J] . Bulletin of the Veterinary Institute in Pulawy,2002,46（2）: 295-299.

[103] 杨金,章强强. 无绿藻的微生物学特性和菌种鉴定 [J]. 中国真菌学杂志,2018,13（2）: 95-100.

[104] CHENGAPPA M M, MADDUX R L, GREER S C, et al.Isolation and identification of yeasts and yeast like organisms from clinical veterinary sources[J].Journal of Clinical Microbiology,1984,19（3）: 427-428.

[105] MORACE G, SANGUINETTI M, POSTERARO B, et al.Identification of various medically important Candida species in clinical specimens by PCR-restriction enzyme analysis[J].Journal of Clinical Microbiology, 1997,35（3）: 667-672.

[106] PORE R S, SHAH AN T A, PORE M D, et al . Occurrence of Prototheca zopfii, a mastitis pathogen, in milk[J] . veterinary microbiology,1987,15（4）: 315-323.

[107] TUBAKI K, SONEDA M.Cultural and taxonomic studies on Prototheca[J].Nagaoa Journal of Medical Science,1959,6: 25-34.

[108] ROESLER U, MOLLER A, HENSEL A, et al.Diversity within the current algal species Prototheca zopf ii: A proposal for two Prototheca zopfii genotypes and description of 10 a novel species, Prototheca blaschkea e sp . nov[J] . International Journal of Systematic And Evolutionary Microbiology,2006,56（6）: 1419-1425.

[109] 王秋东,刘琪,董志民,等.内蒙古部分地区致奶牛乳房炎金黄色葡萄球菌和产色葡萄球菌的分子流行病学及耐药性研究[J].中国预防兽医学报,2017,39(4):5.

[110] MARQUES S, SILVA E, KRAFT C, et al.Bovine mastitis associated with Prototheca blaschkeae[J].Journal of Clinical Microbiology, 2008, 46(6): 1941–1945.

[111] MARQUES S, HUSS V A R, PFISTERER K, et al. Internal transcribed spacer sequence-based rapid molecular identification of Prototheca zopfii and Prototheca blaschkeae directly from milk of infected cows[J]. Journal of Dairy Science, 2015, 98(5): 3001–3009.

[112] SHAVE C D, MILLYARD L, MAY R C .Now for something completely different: Prototheca, pathogenic algae[J].PLoS Pathogens, 2021, 17(4): e1009362.

[113] SUDMAN M S, MAJKA J A, KAPLAN W.Primary mucocutancous protothecosis in a dog[J].J Am Journal of the American Veterinary Medical Association, 1978, 163: 1372–474.

[114] BLASCHKE-HELLMESSEN R, SCHUSTER H, BERGMANN V.Differenzierung von Varianten bei Prototheca zopfii (Kruger 1894)[J] . Arch Exp Veterinarmed, 1985, 39: 387–397.

[115] ROESLER U, SCHOLZ H, HENSEL A . Emended phenotypic characterization of Prototheca zopfii: A proposal for three biotypes and standards for their identification[J]. International Journal of Systematic and Evolutionary Microbiology, 2003, 53(4): 1195–1199.

[116] AALBAEK B, JENSEN H E, HUDA A.Identification of Prototheca from bovine mastitis in Denmark[J].APMIS, 1998, 106(4): 483–488.

[117] ROESLER U, HENSEL A . Eradication of Prototheca zopfii infection in a dairy cattle herd[J].DTW Deutsche Tierarztliche Wothenschrift, 2003, 110(9): 374–377.

[118] Muhammad Shahid. 基因Ⅱ型 Prototheca zopfii 感染奶牛乳腺上皮细胞和小鼠模型的免疫病理学研究[D]. 北京:中国农业大学,2019.

[119] SHAHID M, COBO E R, CHEN L B, et al.Prototheca zopfii genotype Ⅱ induces mitochondrial apoptosis in models of bovine mastitis[J].Scientific Reports, 2020, 10: 698.

[120] SHAHID M, GAO J, ZHOU Y N, et al.Prototheca zopfii isolated from bovine mastitis induced oxidative stress and apoptosis in bovine mammary epithelial cells[J].Oncotarget,2017,8（19）: 31938-31947.

[121] LERCHE M.Eine durch Algen（Prototheca）hervorgerufene mastitis der Kuh[J].Berl Munch Tierarztl Wochenschr,1952,4: 64-69.

[122] JÁNOS I S, RÁTZ F, SZIGETI G, et al . Review of the microbiological, pathological, and clinical aspects of bovine mastitis caused by the alga Prototheca zopfii[J].The Veterinary Quarterly, 2001,23（2）: 58-61.

[123] LAGNEAU, P E.First isolation of Prototheca zopfii in bovine mastitis in Belgium[J].Journal de Mycologie Medicale,1996,6: 145- 148.

[124] BUZZINI P, TURCHETTI B, FACELLI R, et al.First large-scale isolation of Prototheca zopfii from milk produced by dairy herds in Italy[J].Mycopathologia,2004,158（4）: 427-430.

[125] MOLLER A, TRUYEN U, ROESLER U.Prototheca zopfii genotype 2: The causative agent of bovine protothecal ma stitis?[J].Veterinary Microbiology,2007,120（3/4）: 370-374.

[126] PIEPER L, GODKIN A, ROESLER U, et al.Herd characteristics and cow-level factors associated with Prototheca mastitis on dairy farms in Ontario, Canada[J]. Journal of Dairy Science,2012,95（10）: 5635-5644.

[127] TOYOTOME T, MATSUI S.Analysis of Prototheca and yeast species isolated from bulk tank milk collected in Tokachi District, Japan [J].Journal of Dairy Science,2022,105（10）: 8364-8370.

[128] GAO J, HOU R G, ZHANG H Q, et al.A novel DNA extraction and duplex polymerase chain reaction assay for the rapid detection of Protothe ca zopfii genotype 2 in milk[J] . Letters in Applied Microbiology,2011,53（3）: 278-282.

[129] SHAVE C D, MILLYARD L, MAY R C .Now for something completely different: Prototheca, pathogenic algae[J].PLoS Pathogens, 2021,17（4）: e1009362.

[130] THORNSBERRY C, BAKER C N, FACKLAM R R. Antibiotic susceptibility of Streptococcus bovis and other group D streptococci causing endocarditis[J].Antimicrobial Agents and Chemotherapy, 1974,5(3): 228-233.

[131] NOBLE C J.Carriage of group D streptococci in the human bowel[J]. Journal of Clinical Pathology,1978,31(12): 1182-1186.

[132] GONZLEZ-QUINTELA A, MARTINEZ-REY C, CASTROAGUDIN J F, et al.Prevalence of liver disease in patients with Streptococcus bovis bacteraemia[J].The Journal of Infection, 2001,42(2): 116-119.

[133] POYART C, QUESNE G, TRIEU-CUOT P.Taxonomic dissection of the Streptococcus bovis group by analysis of manganese-dependent superoxide di smut as e gene (sodA) sequences: Reclassification of 'Streptococcus infantarius subsp.coli' as Streptococcus lutetiensis sp.nov. and of Streptococcus bovis biotype 11.2 as Streptococcus pasteurianus sp.nov[J].International Journal of Systematic and Evolutionary Microbiology,2002,52(Pt 4): 1247-1255.

[134] DAGAN R, SHRIKER O, HAZAN I, et al.Prospective study to determine clinical relevance of detection of pneumococcal DNA in sera of children by PCR[J].Journal of Clinical Microbiology,1998, 36(3): 669-673.

[135] METTANAND A S, KAMALANATHAN P, DHANANJAN AMALIE K.Streptococcus bovis-unusual etiology of meningitis in a neonate with Down syndrome: A case report[J].Journal of Medical Case Reports,2018,12(1): 93.

[136] CORREDOIRA J, ALONSO M P, COIRA A, et al.Characteristics of Streptococcus bovis endocarditis and its differences with Streptococcus viridans endocarditis[J].European Journal of Clinical Microbiology and Infectious Diseases,2008,27(4): 285-291.

[137] DURANTE-MANGONI E, BRADLEY S, SELTON-SUTY C, et al.Current features of infective endocarditis in elderly patients: Results of the International Collaboration on Endocarditis Prospective Cohort Study[J].Archives of Internal Medicine,2008,

168（19）：2095-2103.

[138] HEADINGS D L, HERRERA A, MAZZI E, et al.Fulminant neonatal septicemia caused by Streptococcus bovis[J].The Journal of Pediatrics,1978,92（2）：282-283.

[139] WHITE B A, LABHSETWAR S A, MIAN A N.Streptococcus bovis bacteremia and fetal death[J].Obstetrics and Gynecology,2002,100（5Pt 2）：1126-1129.

[140] JIN D, CHEN C, LI L Q, et al.Dynamics of fecal microbial communities in children with diarrhea of unknown etiology and genomic analysis of associated Streptococcus lutetien sis[J].BMC Microbiology,2013,13：141.

[141] HOLMES A R, MCNAB R, MILLSAP K W, et al.The pavA gene of Streptococcus pneumoniae encodes a fibronectin-binding protein that is essential for virulence[J].Molecular Microbiology,2001,41（6）：1395-1408.

[142] SPELLERBERG B, ROZDZINSKI E, MARTIN S, et al.Lmb, a protein with similarities to the LraI adhesin family, mediates attachment of Streptococcus agalactiae to human laminin[J].Infection and Immunity,1999,67（2）：871-878.

[143] BERES S B, SYLVA G L, BARBIAN K D, et al.Genome sequence of a serotype M3 strain of group A Streptococcus：Phage-encoded toxins, the high-virulence phenotype, and clone emergence[J].Proceedings of the National Academy of Sciences of the United States of America,2002,99（15）：10078-10083.

[144] TAKAMATSU D, NISHINO H, ISHIJI T, et al.Genetic organization and preferential distribution of putative pilus gene clusters in Streptococcus suis[J].Veterinary Microbiology,2009,138（1/2）：132-139.

[145] RUSNIOK C, COUVÉ E, DA CUNHA V, et al.Genome sequence of Streptococcus gallolyticus：insights into its adaptation to the bovine rumen and its ability to cause endocarditis[J].Journal of Bacteriology,2010,192（8）：2266-2276.

[146] RAJAM G, PHILLIPS D J, WHITE E, et al.A functional epitope of

the pneumococcal surface adhesin A activates nasopharyngeal cells and increases bacterial internalization[J].Microbial Pathogenesis, 2008,44（3）: 186-196.

[147] TETTELIN H, MASIGNANI V, CIESLEWICZ M J, et al.Complete genome sequence and comparative genomic analysis of an emerging human pathogen, serotype V Streptococcus agalactiae[J]. Proceedings of the National Academy of Sciences of the United States of America,2002,99（19）: 12391-12396.

[148] LI N I H, LIU T T, TEN G Y T, et al. Sequencing and comparative genome analysis of two pathogenic Streptococcus gallolyticus subspecies: Genome plasticity, adaptation and virulence[J]. PLoS One,2011,6（5）: e20519.

[149] BERES S B, RICHTER E W, NAGIEC M J, et al.Molecular genetic anatomy of inter-and intraserotype variation in the human bacterial pathogen group A Streptococcus[J].Proceedings of the National Academy of Sciences of the United States of America,2006,103（18）: 7059-7064.

[150] ALMUZARA M, BONOFIGLIO L, CITTADINI R, et al.First case of Streptococcus lutetiensis bacteremia involving a clindamycin-resistant isolate carrying the lnuB gene[J].Journal of Clinical Microbiology,2013,51（12）: 4259-4261.

[151] PIVA S, PIETRA M, SERRAINO A, et al.First description of Streptococcus lutetiensis from a diseased cat[J].Letters in Applied Microbiology, 2019,69（2）: 96-99.

[152] RANTALA S. Streptococcus dysgalactiae subsp. equisimilis bacteremia: An emerging infection[J].European Journal of Clinical Microbiology and Infectious Diseases, 2014,33（8）: 1303-1310.

[153] ZHANG S Y, PIEPERS S, SHAN R X, et al.Phenotypic and genotypic characterization of antimicrobial resistance profiles in Streptococcus dysgalactiae isolated from bovine clinical mastitis in 5 provinces of China[J].Journal of Dairy Science,2018,101（4）: 3344-3355.

[154] NISHIKI I, YOSHIDA T, FUJIWARA A. Complete genome

sequence and characterization of virulence genes in Lancefield group C Streptococcus dysgalactiae isolated from farmed amberjack (Seriola dumerili)[J].Microbiology and Immunology,2019,63(7):243-250.

[155] MORENO L Z, DA COSTA B L P, MATAJIRA C E C, et al.Molecular and antimicrobial susceptibility profiling of Streptococcus dysgalactiae isolated from swine[J].Diagnostic Microbiology and Infectious Disease,2016,86(2):178-180.

[156] JORDAL S, GLAMBEK M, OPPEGAAR D O, et al.New tricks from an old cow: Infective endocarditis caused by Streptococcus dysgalactiae subsp. dysgalactiae[J].Journal of Clinical Microbiology,2015,53(2):731-734.

[157] 陆承平.兽医微生物学[M].4版.北京:中国农业出版社,2007.

[158] 东秀珠,蔡妙英.常见细菌系统鉴定手册[M].北京:科学出版社,2001.

[159] LIU B, ZHENG D, JIN Q, et al.VFDB 2019:A comparative pathogenomic platform with an interactive web interface[J].Nucleic Acids Research,2019,47(D1):D687-D692.

[160] ALVES-BARROCO C, ROMA-RODRIGUES C, BALA-SUBRAMANIAN N, et al.Biofilm development and computational screening for new putative inhibitors of a homolog of the regulatory protein BrpA in Streptococcus dysgalactiae subsp.dysgalactiae[J].International Journal of Medical Microbiology,2019,309(3/4):169-181.

[161] O'HALLORAN F, BEECHER C, CHAURIN V, et al. Lactoferrin affects the adherence and invasion of Streptococcus dysgalactiae ssp. dysgalactiae in mammary epithelial cells[J].Journal of Dairy Science,2016,99(6):4619-4628.

[162] MADUREIRA P, BAPTIST A M, VIEIRA M, et al. Streptococcus agalactiae GAPDH is a virulence-associated immunomodulatory protein[J].Journal of Immunology,2007,178(3):1379-1387.

[163] PEREZ-CASAL J, POTTER A A. Glyceradehyde-3-phosphate dehydrogenase as a suitable vaccine candidate for protection against

bacterial and parasitic diseases[J].Vaccine,2016,34（8）: 1012-1017.

[164] FONTAINE M C, PEREZ-CASAL J, SONG X M, et al.Immunisation of dairy cattle with recombinant Streptococcus uberis GapC or a chimeric CAMP antigen confers protection against het ero logous bacterial challenge[J].Vaccine,2002,20（17/18）: 2278-2286.

[165] CHOU C C, LIN M C, SU F J, et al.Mutation in cyl operon alters hemolytic phenotypes of streptococcus agalactiae[J].Infection, Genetics and Evolution,2019,67: 234-243.

[166] JONSSON H, FRYKBERG L, RANTAMÄKI L, et al.MAG, a novel plasma protein receptor from Streptococcus dysgalactiae[J]. Gene, 1994,143（1）: 85-89.

[167] SONG X M, PEREZ-CASAL J, FONTAINE M C, et al.Bovine immunoglobulin A（IgA）-binding activities of the surface-expressed Mig protein of Streptococcus dysgalactiae[J]. Microbiology, 2002,148（Pt 7）: 2055-2064.

[168] GAO J, BARKEMA H W, ZHANG L M, et al.Incidence of clinical mastitis and distribution of pathogens on large Chinese dairy farms[J] . Journal of Dairy Science,2017,100（6）: 4797-4806.

[169] BI Y, WANG Y J, QIN Y, et al.Prevalence of bovine mastitis pathogens in bulk tank milk in China[J].PLo S One,2016,11（5）: e0155621 .

[170] HEIKKILÄ A M, LISKI E, PYÖRÄLÄ S, et al.Pathogen-specific production losses in bovine mastitis[J].Journal of Dairy Science, 2018, 101（10）: 9493-9504.

[171] BOTREL M A, HAENNI M, MORIGNAT E, et al.Distribution and antimicrobial resistance of clinical and subclinical mastitis pathogens in dairy cows in Rhône-Alpes, France[J].Foodborne Pathogens and Disease,2010, 7（5）: 479-487.

[172] CAMERON M, SAAB M, HEIDER L, et al . Antimicrobial susceptibility patterns of environmental streptococci recovered from bovine milk samples in the maritime provinces of Canada[J]. Frontiers in Veterinary Science,2016,3: 79.

[173] SOLTAU J B, EINAXE, KLENGEL K, et al. Within-herd prevalence thresholds for herd-level detection of mastitis pathogens using multiplex real-time PCR in bulk tank milk samples[J].Journal of dairy science,2017,100（10）：8287-8295.

[174] VAKKAMAKI J, TAPONEN S, HEIKKILA A M, et al.Bacteriological etiology and treatment of mastitis in Finnish dairy herds[J].Acta Veterinaria Scandinavica,2017,59（1）：33.

[175] SHUM L W C, MCCONNEL C S, GUNN A A, et al.Environmental mast it is in intensive high-producing dairy herds in New South Wales[J].Australian Veterinary Journal,2009,87（12）：469-475.

[176] RUEGG P L, OLIVEIRA L, JIN W, et al.Phenotypic antimicrobial susceptibility and occurrence of selected resistance genes in gram-positive mastitis pathogens isolated from Wisconsin dairy cows[J]. Journal of Dairy Science,2015,98（7）：4521-4534.

[177] COBO-ÁNGEL C, JARAMILLO-JARAMILLO A S, LASSO-ROJAS L M, et al.Streptococcus agalactiae is not always an obligate intramammary pathogen: Molecular epidemiology of GBS from milk, feces and environment in Colombian dairy herds[J].PLoS One,2018,13（12）：e0208990.

[178] MOROZUMI M, WAJIM A T, KUWATA Y, et al.Associations between capsular serotype, multilocus sequence type, and macrolide resistance in Streptococcus agalactiae isolates from Japanese infants with invasive infections[J]. Epidemiology and infection,2014,142（4）：812-819.

[179] BRIMIL N, BARTHELL E, HEINDRICH S U, et al.Epidemiology of Streptococcus agalactiae colonization in Germany[J].International Journal of Medical Microbiology,2006,296（1）：39-44.

[180] 朱战波,王红,刘宇,等.奶牛乳房炎链球菌分离株的鉴定及药敏试验[J].现代畜牧兽医,2007（12）：8-11.

[181] KEEFE G P.Streptococcus agalactiae mastitis: A review[J].The Canadian Veterinary Journal,1997,38（7）：429-437.

[182] 侯小露.奶牛乳房炎链球菌的分离鉴定及其多克隆抗体的制备[D].南宁：广西大学,2013.

[183] 曹授俊,蔡泽川,姜小平.北京某地区奶牛隐性乳房炎病原菌的分离鉴定及药敏试验[J].山东畜牧兽医,2012,4:7-9.

[184] 高桂生,高光平,邵新华,等.冀东地区奶牛隐性乳房炎病原菌的分离鉴定[J].中兽医医药杂志,2012,4:17-19.

[185] 曲家华,冯万宇,周庆民.黑龙江西部地区奶牛临床型乳房炎病原菌分离鉴定及耐药性分析[J].中国草食动物科学,2012,32(5):53-54.

[186] 沈永聪,金云行,沈祥广,等.奶牛隐性乳房炎发病率调查及病原菌的分离鉴定[J].中国兽医杂志,2012,48(5):48-49.

[187] 王东,李发万,王旭青,等.宁夏部分地区奶牛乳腺炎病原菌的分离鉴定[J].动物医学进展,2012,33:119-122.

[188] 王旭荣,李宏胜,李建喜,等.奶牛临床型乳房炎的细菌分离鉴定与耐药性分析[J].中国畜牧兽医,2012,39(7):195-198.

[189] 黄瑛,尹晓敏,李丽好,等.奶牛临床型乳房炎病原菌的分离与鉴定[J].中国畜牧兽医,2007,34(12):74-76.

[190] 殷波,熊焰.临床型奶牛乳房炎病原分离鉴定及药敏试验[J].黑龙江畜牧兽医,2006,2:61-62.

[191] 袁永隆,张永欣.我国奶牛房腺炎常见病原菌的区系调查[J].中国农业科学,1992,25(4):70-76.

[192] 李宏胜,郁杰,李新圃,等.奶牛乳房炎类型与病原菌感染之间相关性的研究[J].动物医学进展,2004,25(3):80-84.

[193] 马保臣,牛家华,高玉君,等.奶牛乳房炎的细菌学研究[J].家畜生态学报,2005,26:35-40.

[194] 许丹宁,侯振中,樊卫东.奶牛隐性乳房炎的流行病学调查及病原菌的分离鉴定[J].现代畜牧兽医,2005,9:40-41.

[195] ARAN A, WEINER K, LIN L, et al.Post-streptococcal auto-antibodies inhibit protein disulfide isomerase and are associated with insulin resistance[J].PLoS One,2010,5(9):e12875.

[196] ABERA M, HABTE T, ARAGAW K, et al.Major causes of mastitis and associated risk factors in smallholder dairy farms in and around Hawassa, Southern Ethiopia[J].Tropical Animal Health and Production,2012,44(6):1175-1179.

[197] SENTITULA YADAV B R, KUMAR R. Incidence of staphylococci

and streptococci during winter in mastitic milk of sahiwal cow and murrah buffaloes[J].Indian Journal of Microbiology,2012,52（2）: 153-159.

[198] FRANCOZ D, BERGERON L, NADEAU M, et al.Prevalence of contagious mastitis pathogens in bulk tank milk in Québec[J]. The Canadian Veterinary Journal = La Revue Veterinaire Canadienne, 2012, 53（10）: 1071-1078.

[199] KATHOLM J, BENNEDSGAARD T W, KOSKINEN M T, et al.Quality of bulk tank milk samples from Danish dairy herds based on realtime polymerase chain reaction identification of mastitis pathogens[J]. Journal of Dairy Science,2012,95（10）: 5702-5708.

[200] CARVALHO-CASTRO G A, SILVA J R, PAIVA L V, et al.Molecular epidemiology of Streptococcus agalactiae isolated from mastitis in Brazilian dairy herds[J].Brazilian Journal of Microbiology: [publication of the Brazilian Society for Microbiology],2017,48（3）: 551-559.

[201] AZEVEDO C, PACHECO D, SOARES L, et al.Prevalence of bovine milk pathogens in Azore an pastures: Mobile versus fixed milking machines[J].Veterinary Record Open,2016,3（1）: e000181.

[202] TENHAGEN B A, KÖSTER G, WALLMANN J, et al.Prevalence of mastitis pathogens and their resistance against antimicrobial agents in dairy cows in Brandenburg, Germany[J].Journal of Dairy Science, 2006,89（7）: 2542-2551.

[203] BI Y L, WANG Y J, QIN Y, et al.Prevalence of bovine mastitis pathogens in bulk tank milk in China[J].PLoS One, 2016,11（5）: e0155621.

[204] NOCARD E.Note sur la maladie des boeufs de la Guadeloupe connue sous le nom de farcin[J].Ann Inst Pasteur,1888,2: 293-302.

[205] EPPINGER H, UEBER EINE NEUE.Pathogene cladothrix und eine durch sie hervorgerufene pseudo-tuberculosis[M].Cladothrichica, 1890.

[206] BLANCHARD R . Parasites véget aux à l'exclusion des bactéries[J].Traite de Pathologie Generale,1896,2: 811-932.

[207] GORDON R E, BARNETT D A, HANDERHAN J E, et al.Nocardia coeliaca, nocardia autotrophica, and the nocardin strain[J]. International Journal of Systematic Bacteriology, 1974, 24（1）: 54-63.

[208] MALDONADO L, HOOKEY J V, WARD A C, et al.The Nocardia salmonicida clade, including descriptions of Nocardia cummidelens sp.nov., Nocardia fluminea sp.nov.and Nocardia soli sp.nov[J]. Antonie Van Leeuwenhoek, 2000, 78（3）: 367-377.

[209] SHIMAHARA Y, HUANG Y F, TSAI M A, et al.Genotypic and phenotypic analysis of fish pathogen, Nocardia seriolae, isolated in Taiwan[J].Aquaculture, 2009, 294（3/4）: 165-171.

[210] KAGEYAMA A, YAZAWA K, ISHIKAWA J, et al .No car dial infections in Japan from 1992 to 2001, including the first report of infection by nocardia transvalensis[J] . European Journal of Epidemiology, 2004, 19（4）: 383-389.

[211] BROWN-ELLIOTT B A, BROWN J M, CONVILLE P S, et al.Clinical and laboratory features of the Nocardia spp.based on current molecular taxonomy[J].Clinical Microbiology Reviews, 2006, 19（2）: 259-282.

[212] MURICY E C M, LEMES R A, BOMBARDA S, et al.Differentiation between nocardia spp.and mycobacterium spp.: Critical aspects for bacteriological diagnosis[J].Revista Do Instituto De Medicina Tropical De São Paulo, 2014, 56（5）: 397-401.

[213] SCHLABERG R, HUARD R C, DELLA-LATTA P. No car dia cyriacigeorgica, an emerging pathogen in the United States[J]. Journal of Clinical Microbiology, 2008, 46（1）: 265-273.

[214] BAFGHI M F, HEIDARIEH P, SOORI T, et al.Nocardia isolation from clinical samples with the paraffin baiting technique[J].Germs, 2015, 5（1）: 12-16.

[215] MCNEIL M M, BROWN J M . The medically important aerobic actinomycetes: Epidemiology and microbiology[J]. Clinical Microbiology Reviews, 1994, 7（3）: 357-417.

[216] BAFGHI M F, HEIDARIEH P, SOORI T, et al.Nocardia isolation

from clinical samples with the paraffin baiting technique[J].Germs, 2015,5（1）: 12-16.

[217] WAUTER S G, AVESAN I V, CHARLIER J, et al . Distribution of nocardia species in clinical samples and their routine rapid identification in the laboratory[J].Journal of Clinical Microbiology, 2005,43（6）: 2624-2628.

[218] KISKA D L, HICKS K, PETTIT D J.Identification of medically relevant Nocardia species with an abbreviated battery of tests[J].Journal of Clinical Microbiology,2002,40（4）: 1346-1351.

[219] VERA-CABRERA L, ORTIZ-LOPEZ R, ELIZONDO-GONZALEZ R, et al . Complete genome sequence analysis of Nocardia brasiliensis HUJEG-1 reveals a saprobic lifestyle and the genes needed for human pathogenesis[J].PLoS One,2013,8（6）: e65425 .

[220] EROKSUZ Y, GURSOY N C, KARAPINAR T, et al . Systemic nocardiosis in a dog caused by Nocardia cyriacigeorgica[J]. BMC Veterinary Research,2017,13（1）: 30.

[221] BHANA S A, TSITSI J M L, et al.Thyrotoxicosis followed by Hypothyroidism due to Suppurative Thyroiditis Caused by No car dia brasiliensis in a Patient with Advanced Acquired Immunodeficiency Syndrome[J].European Thyroid Journal,2014,3（1）: 65-68.

[222] FOX L K.Mycoplasma mastitis: Causes, transmission, and control[J].The Veterinary Clinics of North America Food Animal Practice,2012,28（2）: 225-237.

[223] RUEGG P.Managing mastitis and producing quality milk[J] .Dairy Production Medicine,2011,10: 207-232.

[224] BAIRD S C, CARMAN J, DINSMORE R P, et al.Detection and identification of Mycoplasma from bovine mastitis infections using a nested polymerase chain reaction[J].Journal of Veterinary Diagnostic Investigation,1999,11: 432-435.

[225] FOX L K, HANCOCK D D, MICKELSON A, et al.Bulk tank milk analysis: Factors associated with appearance of Mycoplasma sp. in milk[J].Journal of Veterinary Medicine B, Infectious Diseases and

Veterinary Public Health,2003,50（5）:235-240.

[226] OLDE RIEKERINK R G M,BARKEMA H W,VEENSTRA S,et al.Prevalence of contagious mastitis pathogens in bulk tank milk in Prince Edward Island[J].The Canadian Veterinary Journal,2006,47（6）:567-572.

[227] PASSCHYN P,PIEPERS S,DE MEULEMEESTER L,et al.Between- herd prevalence of Mycoplasma bovis in bulk milk in Flanders,Belgium[J].Research in Veterinary Science,2012,92(2):219-220.

[228] PARKER A M,HOUSE J K,HAZELTON M S,et al.Bulk tank milk antibody ELISA as a biosecurity tool for detecting dairy herds with past exposure to Mycoplasma bovis[J].Journal of Dairy Science,2017,100（10）:8296-8309.

[229] ROSALES R S,CHURCHWARD C P,SCHNEE C,et al . Global multilocus sequence typing analysis of Mycoplasma bovis isolates reveals two main population clusters[J] . Journal of Clinical Microbiology,2015,53（3）:789-794.

[230] TANNER A C,WU C C.Adaptation of the Sensititre[registered trademark] broth microdilution technique to antimicrobial susceptibility testing of mycoplasma gallisepticum[J].Avian Diseases,1992,36（3）:714-717.

[231] FOX L K,KIRK J H,BRITTEN A.My coplasma mastitis:A review of transmission and control[J].Journal of Veterinary Medicine B,Infectious Diseases and Veterinary Public Health,2005,52（4）:153-160.

[232] MAUNSELL F P,WOOLUMS A R,FRANCOZ D,et al.Mycoplasma bovis infections in cattle[J].Journal of Veterinary Internal Medicine,2011,25（4）:772-783.

[233] SONG X B,HUANG X P,XU H Y,et al.The prevalence of pathogens causing bovine mastitis and their associated risk factors in 15 large dairy farms in China:An observational study[J].Veterinary Microbiology,2020,247:108757.

[234] GHAZAEI C . My coplasmal mastitis in dairy cows in the Moghan

region of Ardabil State, Iran[J].Journal of the South African Veterinary Association,2006,77（4）:222-223.

[235] PASSCHYN P, PIEPERS S, DE MEULEMEESTER L, et al.Between- herd prevalence of Mycoplasma bovis in bulk milk in Flanders, Belgium[J].Research in Veterinary Science,2012,92(2):219-220.

[236] FILIOUSS IS G, CHRISTODOULOPOUL OS G, THATCHER A, et al.Isolation of Mycoplasma bovis from bovine clinical mastitis cases in Northern Greece[J].Veterinary Journal,2007,173（1）:215-218.

[237] LIU Y, XU S Y, LI M Y, et al.Molecular characteristics and antibiotic susceptibility profiles of Mycoplasma bovis associated with mastitis on dairy farms in China[J].Preventive Veterinary Medicine, 2020,182:105106.

[238] MENGHWAR H, HE C F, ZHANG H, et al.Genotype distribution of Chinese Mycoplasma bovis isolates and their evolutionary relationship to strains from other countries[J].Microbial Pathogenesis,2017,111: 108-117.

[239] 温靖,李丹,郭婷,等.中国奶牛牛支原体流行病学调查分析[J].中国农业大学学报,2022,27:75-82.

[240] 李媛,闫磊,刘桐,等.中国西部和北部边境4省区牛支原体血清学流行病学调查及分析[J].中国预防兽医学报,2021,43:717-721.

[241] MENGHWAR H, HE C F, ZHANG H, et al.Genotype distribution of Chinese Mycoplasma bovis isolates and their evolutionary relationship to strains from other countries[J].Microbial Pathogenesis,2017,111: 108-117.

[242] FOX L K.Mycoplasma mastitis: Causes, transmission, and control[J].The Veterinary Clinics of North America Food Animal Practice, 2012,28（2）:225-237.

[243] FOX L K, HANCOCK D D, MICKELSON A, et al.Bulk tank milk analysis: Factors associated with appearance of Mycoplasma sp. in milk[J].Journal of Veterinary Medicine.B, Infectious Diseases and Veterinary Public Health,2003,50（5）:235-240.

[244] LYSNYANSKY I, FREED M, ROSALES R S, et al.An overview

of Mycoplasma bovis mastitis in Israel（2004-2014）[J].Veterinary Journal,2016,207: 180-183.

[245] PUNYAPORNWITHAYA V, FOX L K, HANCOCK D D, et al.Time to clearance of mycoplasma mastitis: The effect of management factors including milking time hygiene and preferential culling[J].The Canadian Veterinary Journal = La Revue Veterinaire Canadienne,2012, 53（10）: 1119-1122.

[246] OTTER A, WRIGHT T, LEONARD D, et al.Mycoplasma bovis mastitis in dry dairy cows[J].The Veterinary Record,2015,177（23）: 601-602.

[247] MENGHWAR H, HE C F, ZHANG H, et al.Genotype distribution of Chinese Mycoplasma bovis isolates and their evolutionary relationship to strains from other countries[J].Microbial Pathogenesis,2017,111: 108-117.

[248] SCHLEIFER K H, KRAUS J, DVORAK C, et al . Transfer of streptococcus lactis and related streptococci to the genus lactococcus gen.nov[J].Systematic and Applied Microbiology,1985,6（2）: 183-195.

[249] KIMURA H, KUSUDA R . Studies on the pathogenesis of streptococcal infection in cultured yellowtails, Seriola spp.: Effect of crude exotoxin fractions from cell - free culture on experimental streptococcal infection[J].Journal of Fish Diseases,1982,5（6）: 471-478.

[250] SHAHI N, MALLIK S K, SAHOO M, et al . First report on characterization and pathogenicity study of emerging Lactococcus garvieae infection in farmed rainbow trout, Oncorhynchus mykiss（Walbaum）, from India[J]. Transboundary and Emerging Diseases, 2018,65（4）: 1039-1048.

[251] COLLINS M D, FARROW J A E, PHILLIPS B A, et al.Streptococcus garvieae sp.nov. and streptococcus plantarum sp.nov[J].Microbiology,1983,129（11）: 3427-3431.

[252] TEIXEIRA L M, MERQUIOR V L, VIANNI M C, et al.Phenotypic and genotypic characterization of atypical Lactococcus garvieae strains isolated from water buffalos with subclinical mastitis and

confirmation of L . garvieae as a senior subjective synonym of Entero-coccus seriolicida[J].International Journal of Systematic Bacteriology,1996,46（3）: 664-668.

[253] DEVRIESE L A, HOMMEZ J, LAEVENS H, et al.Identification of aesculin -hydrolyzing streptococci, lactococci, aero cocci and enterococci from subclinical intramammary infections in dairy cows[J] . Veterinary Microbiology,1999,70（1/2）: 87-94.

[254] RODRIGUES M X, LIMA S F, HIGGINS C H, et al . The Lactococcus genus as a potential emerging mastitis pathogen group: A report on an outbreak investigation[J].Journal of Dairy Science, 2016, 99（12）: 9864-9874.

[255] ERACLIO G, RICCI G, MORONI P, et al.Sand bedding as a reservoir for Lactococcus garvieae dissemination in dairy farms[J] . Canadian Journal of Microbiology,2019,65（1）: 84-89.

[256] RASMUSSEN M.Aerococci and aerococcal infections[J].The Journal of Infection,2013,66（6）: 467-474.

[257] LIU G, LIU Y X, ALI T, et al . Molecular and phenotypic characterization of aerococcus viridan s associated with subclinical bovine mastitis[J].PLoS One,2015,10（4）: e0125001.

[258] WILLIAMS R E O, HIRCH A, COWAN S T.Aerococcus, a new bacterial genus[J].Journal of General Microbiology,1953,8（3）: 475-480.

[259] 刘钢. 奶牛乳房炎源性浅绿气球菌的表型、基因型和致病机制研究[D]. 北京: 中国农业大学,2017.

[260] WILLIAMS R E, HIRCH A, COWAN S T.Aerococcus, a new bacterial genus[J].Journal of General Microbiology,1953,8（3）: 475-480.

[261] KERBAUGH M A, EVANS J B. Aerococcus viridans in the hospital environment[J]. Applied Microbiology,1968,16（3）: 519-523.

[262] SAISHU N, MORIMOTO K, YAMASATO H, et al.Characterization of Aerococcus viridan s isolated from milk samples from cows with mastitis and manure samples[J].The Journal of Veterinary Medical Science,2015,77（9）: 1037 - 1042.

[263] GREENWOOD S J, KEITH I R, DESPRES B M, et al. Genetic characterization of the lobster pathogen Aerococcus viridans var. homari by 16S rRNA gene sequence and RAPD[J].Diseases of Aquatic Organisms,2005,63（2/3）: 237-246.

[264] KE X L, LU M X, YE X, et al . Recovery and pathogenicity analysis of Aerococcus viridans isolated from tilapia（Orecohromis niloticus）cultured in southwest of China[J].Aquaculture,2012, 342/343: 18-23.

[265] DEVRIESE L A, HOMMEZ J, LAEVENS H, et al.Identification of aesculin -hydrolyzing streptococci, lactococci, aerococci and enterococci from subclinical intramammary infections in dairy cows[J] . Veterinary Microbiology,1999,70（1/2）: 87-94.

[266] BOSSHARD P P, ABELS S, ALTWEGG M, et al.Comparison of conventional and molecular methods for identification of aerobic catalase-negative gram-positive cocci in the clinical laboratory[J]. Journal of Clinical Microbiology,2004,42（5）: 2065-2073.

[267] PITKÄLÄ A, HAVERI M, PYÖRÄLÄ S, et al.Bovine mastitis in Finland 2001-prevalence, distribution of bacteria, and antimicrobial resistance[J].Journal of Dairy Science,2004,87（8）: 2433-2441.

[268] ZADOKS R N, GONZALEZ R N, BOOR K J, et al.Mastitis-causing streptococci are important contributors to bacterial counts in raw bulk tank milk[J].Journal of Food Protection,2004,67（12）: 2644-2650.

[269] ŠPAKOVÁ T, ELEEKO J, VASIL M, et al.Limited genetic diversity of Aerococcus viridans strains isolated from clinical and subclinical cases of bovine mastitis in Slovakia[J].Polish Journal of Veterinary Sciences,2012,15（2）: 329-335.

[270] HOGAN J, LARRY SMITH K . Coliform mastitis[J] . Veterinary Research,2003,34（5）: 507-519.

[271] DE VLIEGHER S, LAEVENS H, DEVRIESE L A, et al.Prepartum teat apex colonization with Staphylococcus chromogenes in dairy heifers is associated with low somatic cell count in early lactation[J]. Veterinary Microbiology,2003,92（3）: 245-252.

[272] TAPONEN S, KOORT J, BJÖRKROTH J, et al.Bovine intramammary infections caused by coagulase-negative staphylococci may persist throughout lactation according to amplified fragment length polymorphism-based analysis[J].Journal of Dairy Science,2007,90（7）: 3301-3307.

[273] SCHUKKEN Y H, GONZALES R N, TIKOFSKY L L, et al. CNS mastitis: Nothing to worry about?[J].Veterinary Microbiology, 2009, 134（1/2）: 9-14.

[274] DE VLIEGHER S, FOX L K, PIEPERS S, et al.Invited review: Mastitis in dairy heifers: Nature of the disease, potential impact, prevention, and control[J].Journal of Dairy Science,2012,95（3）: 1025-1040.

[275] TENHAGEN B A, KOSTER G, WALLMANN J, et al.Prevalence of mastitis pathogens and their resistance against antimicrobial agents in dairy cows in Brandenburg, Germany[J].Journal of Dairy Science, 2006,89（7）: 2542-2551.

[276] ROBERSON J R, MIXON J, ROHRBACH B, et al.Etiologic agents associated with high SCC dairy herds[C].In: Proceedings of the 24th World Buiatrics Congress, Nice, France,2006.

[277] DINGWELL R T, LESLIE K E, SCHUKKEN Y H, et al.Association of cow and quarter-level factors at drying-off with new intramammary infections during the dry period[J].Preventive Veterinary Medicine, 2004,63（1/2）: 75-89.

[278] HALTIA L, HONKANEN-BUZALSKI T, SPIRIDONOVA I, et al.A study of bovine mastitis, milking procedures and management practices on 25 Estonian dairy herds[J].Acta Veterinaria Scandinavica,2006,48（1）: 22.

[279] PYÖRÄLÄ S . Treatment of clinical mastitis : local and/ or systemic? Short or long?[C]. In: Proceedings of the 24th World Buiatrics Congress, Nice, France,2006: 250-259.

[280] RASPANTICG, BONETTOCC, VISSIOC, et al.Prevalence and antibiotic susceptibility of coagulase-negative Staphylococcus species from bovine subclinical mastitis in dairy herds in the central

[281] SUPRE K, HAESEBROUCK F, ZADOK S R N, et al. Some coagulase-negative Staphylococcus species affect udder health more than others[J].Journal of Dairy Science,2011,94（5）：2329-2340.

[282] PIESSENS V, VAN COILLIE E, VERBIST B, et al.Distribution of coagulase-negative Staphylococcus species from milk and environment of dairy cows differs between herds[J].Journal of Dairy Science,2011, 94（6）：2933-2944.

[283] DE VLIEGHER S, ZADOKS R N, BARKEMA H W.Heifer and CNS mastitis[J].Veterinary Microbiology,2009,134（1/2）：1-2.

[284] DE VISSCHER A, PIEPERS S, HAESEBROUCK F, et al.Intramammary infection with coagulase-negative staphylococci at parturition：Species-specific prevalence, risk factors, and effect on udder health[J].Journal of Dairy Science,2016,99（8）：6457-6469.

[285] TREMBLAY Y D N, LAMARCHE D, CHEVER P, et al. Characterization of the ability of coagulase-negative staphylococci isolated from the milk of Canadian farms to form biofilms[J].Journal of Dairy Science, 2013,96：234-246.

[286] PYORALA S, TAPONEN S.Coagulase-negative staphylococci-Emerging mastitis pathogens[J].Veterinary Microbiology,2009,134（1/2）：3-8.

[287] THORBERG B M, KÜHN I, AARESTRUP F M, et al.Pheno – and genotyping of Staphylococcus epidermidis isolated from bovine milk and human skin[J].Veterinary Microbiology,2006,115（1/2/3）：163-172.

[288] 曹蕴,李伟,徐葵花．表皮葡萄球菌感染调查及耐药性分析[J]．当代医学,2019,25（21）：39-41.

[289] 余贺．医学微生物学[M]．北京：人民卫生出版社,1983.

[290] 林永焕．临床败血症[M]．西安：陕西科学技术出版社,1998.

[291] 陆德源,任中原．医学微生物学[M].3版．北京：人民卫生出版社,1996.

[292] FOX L K, ZADOKS R N, GASKINS C T. Biofilm production by

Staphylococcus aureus associated with intramammary infection[J]. Veterinary Microbiology,2005,107(3/4): 295-299.

[293] KAUR H, KUMAR P, RAY P, et al .Biofilm formation in clinical isolates of group B streptococci from North India[J].Microbial Pathogenesis,2009,46(6): 321-327.

[294] STEWART P S, COSTERTON J W.Antibiotic resistance of bacteria in biofilms[J].Lancet,2001, 358(9276): 135-138.

[295] DONLAN R M.Role of biofilms in antimicrobial resistance[J]. ASAIO Journal,2000,46(6): S47-S52.

[296] BURKI S, FREY J, PIL O P. Virulence, persistence and dissemination of Mycoplasma bovis[J].Veterinary Microbiology, 2015,179(1/2): 15-22.

[297] FERNANDES J B C, ZANARDO L G, GALV~AO N N, et al. Escherichia coli from clinical mastitis: Serotypes and virulence factors[J].Journal of Veterinary Diagnostic Investigation,2011,23 (6): 1146-1152.

[298] ATULYA M, JESIL MATHEW A, RAO J V, et al.Influence of milk components in establishing bio film mediated bacterial mast it is infections in cattle: A fractional factorial approach[J].Research in Veterinary Science,2014,96(1): 25-27.

[299] MCAULIFFE L, ELLIS R J, MILES K, et al.Biofilm formaction by mycoplasma species and its role in environmental persistence and survival[J].Microbiology,2006,152(4): 913-922.

[300] ORTEGA MORENTE E, FERNANDEZ-FUENTES M A, BURGOS M J, et al.Biocide tolerance in bacteria[J].International Journal of Food Microbiology,2013,162(1): 13-25.

[301] PEARSON J, MACKIE D.Factors associated with the occurrence, cause and outcome of clinical mastitis in dairy cattle[J]. Veterinary Record,1979,105(20): 456-463.

[302] THORBERG B M, KÜHN I, AARESTRUP F M, et al.Pheno - and genotyping of Staphylococcus epidermidis isolated from bovine milk and human skin[J].Veterinary Microbiology,2006,115(1/2/3): 163-172.

[303] 陆承平. 兽医微生物学 [M].5 版. 北京：中国农业出版社, 2013.

[304] 田秋丰, 张红, 尹珺伊, 等.1 株奶牛源松鼠葡萄球菌的分离鉴定及耐药性分析 [J]. 中国畜牧兽医, 2021, 48（04）: 1482-588.

[305] NAZIPI S, VANGKILDE-PEDERSEN S G, BUSCK M M, et al.An antimicrobial Staphylococcus sciuri with broad temperature and salt spectrum isolated from the surface of the African social spider, Stegodyphus dumicola[J].Antonie Van Leeuwenhoek, 2021, 114（3）: 325-335.

[306] STEPANOVIĆ S, DAKIĆ I, MARTEL A, et al .A comparative evaluation of phenotypic and molecular methods in the identification of members of the Staphylococcus sciuri group[J].Systematic and Applied Microbiology, 2005, 28（4）: 353-357.

[307] HU X J, ZHENG B W, JIANG H Y, et al.Draft genome sequence of staphylococcus sciuri subsp. sciuri strain Z8, isolated from human skin[J].Genome Announcements, 2015, 3（4）: e00714-e00715.

[308] DAKIĆ I, MORRISON D, VUKOVIĆ D, et al . Isolation and molecular characterization of Staphylococcus sciuri in the hospital environment [J].Journal of Clinical Microbiology, 2005, 43（6）: 2782-2785.

[309] LU L X, HE K W, NI Y X, et al.Exudative epidermitis of piglets caused by non-toxigenic Staphylococcus sciuri[J].Veterinary Microbiology, 2017, 199: 79-84.

[310] NEMEGHAIR E S, VANDERHAEGHE N W, ARGUDÍN M A, et al. Characterization of methicillin-resistant Staphylococcus sciuri isolates from industrially raised pigs, cattle and broiler chickens[J].Journal of Antimicrobial Chemotherapy, 2014, 69（11）: 2928-2934.

[311] NEYAZ L, KARKI A B, FAKHR M K.Draft genome sequence of megaplasmid-bearing staphylococcus sciuri strain B9-58B, isolated from retail pork[J].Microbiology Resource Announcements, 2020, 9（1）: e01474-e01419.

[312] LI H H, LI X Y, LU Y, et al.Staphylococcus sciuri exfoliative toxin C is a dimer that modulates macrophage functions [J].Canadian Journal of Microbiology, 2011, 57（9）: 722-729.

第三节　牛奶中的体细胞数

哺乳动物均进化出乳房组织,以哺育幼崽。然而,得益于遗传选择以及挤奶技术的进步,乳腺所产生的乳汁,不仅远超犊牛的消耗需求,更是大大超出了乳房组织的容纳量。为提升牛奶产量而对奶牛进行选育,以及运用机器挤奶来排空乳汁,这给牛的乳房带来了非自然的压力,致使这些动物乳房感染的概率上升。为抵御乳腺感染,体细胞会被释放至乳汁中。这些细胞不但能够抵御感染,还能够修复组织损伤。在所有发达国家,乳体细胞计数(体细胞数)均被当作监测奶牛群中乳腺炎患病率的标识,作为评判加工商所收购原料奶质量的指标,同时也是衡量农场牛奶生产卫生状况的更具普遍性的指标。在各类乳质量筛选试验当中,对乳体细胞的估算乃是检测乳腺炎亚临床形式最为有效的手段。在欧盟、中国、新西兰、澳大利亚、瑞士以及加拿大,散装牛奶体细胞数的法定限量为$(3 \sim 4) \times 10^5$个/mL;在南非和巴西为5×10^5个/mL;美国则为7.5×10^5个/mL。在发达国家,细胞计数低的牛奶价钱更高,原因在于此类牛奶的保质期更长。然而,在发展中国家,牛奶的销售依旧基于其脂肪含量。

一、奶牛乳腺

哺乳动物的名字来源于"乳腺"这个词,乳腺是存在于雌性哺乳动物体内的产乳器官。它是一种变异的汗腺,由乳房和乳头组成。雄(男)性和雌(女)性在出生时均未发育,当雌(女)性进入青春期时,乳腺开始发育为第二性征。随着雌性个体首次生育幼崽,乳腺发育到最成熟的阶段,开始分泌乳汁。泌乳乳腺的基本组成部分是排列着分泌乳汁的立方细胞(上皮细胞)的腺泡,它们被肌上皮细胞或肌肉细胞包围,在催产素的作用下这些细胞会收缩,将乳汁从肺泡腔排出到导管中。多个腺泡连接在一起形成称为小叶的结构组,每个小叶都有一个分泌乳汁的管道,这些管道最终汇入乳头的开口。乳汁在腺泡区不断合成,并在两次挤奶之间储存在腺泡、乳管和乳池中。

乳腺有多种类型的细胞,能够参与细胞免疫和体液免疫,有助于防止病原体入侵乳腺。病原体通常在哺乳期之前、期间和之后通过乳头管进入乳腺。在干燥期和挤奶之间,乳管被角蛋白塞密封,角蛋白塞是由乳管内层的鳞状上皮形成的。这形成了一个有效的物理屏障和具有杀微生物特性的防御机制,防止微生物的入侵。然而,如果角蛋白塞受损,乳头管的渗透性会暂时或永久地增加,从而增加乳房感染的机会。一旦发生细菌感染,乳汁体细胞,尤其是乳汁白细胞就会显著增加。

二、牛奶体细胞计数

乳体细胞指的是来源于体内的细胞,通常在牛奶中含量较低。正常乳汁中的乳体细胞大部分是来自乳腺分泌组织的细胞(上皮细胞),还有一些是白细胞。乳头是乳房防止细菌入侵的第一道防御线。在正常情况下,奶牛不挤奶时乳头括约肌处于关闭状态,在挤奶期间,乳头括约肌开张,为细菌浸入乳房打开了通道,特别是当有多余的空气进入挤奶器,如滑脱、二次上杯、取下挤奶器时没有先关掉真空泵,都可能导致奶中或乳头末端的细菌会被吸进乳头管和乳腺池中并在乳腺中繁殖,从而引起乳腺感染。乳体细胞是第二道防线。乳汁中的上皮细胞是腺泡和乳头管的乳腺上皮脱落形成的,这些细胞出现在乳汁中是一种正常的生理现象,是上皮再生所必需的过程。乳中大多数脱落的上皮细胞是有活力的,并表现出完全分化的腺泡细胞的特征。

牛奶中的白细胞属于血源性乳体细胞,是防御系统的一部分,它们能够对抗疾病并帮助修复受损的组织。牛奶体细胞数是指每毫升牛奶中包括巨噬细胞、淋巴细胞、嗜中性白细胞及少量乳腺组织上皮细胞在内的体细胞总数。任何乳腺内感染(如乳腺炎)都会导致乳汁中白细胞的增加,表明所产乳汁的卫生状况不佳。牛奶体细胞计数被量化为每毫升牛奶的细胞数量。当牛奶体细胞计数数量在 10 万左右时,表明该动物未受影响。然而,当牛奶中体细胞数量超过 20 万个 /mL 时,奶牛和水牛至少有四分之一的可能性被感染。正常情况下,牛奶中体细胞数在 20 万 ~ 30 万个 /mL;当乳房产生炎症时,机体将分泌大量白细胞,体细胞数一般超过 50 万个 /mL。随着体细胞数量的增加,它与感染的严重程度和奶牛感染的季度数量直接相关。

三、牛奶中体细胞的释放机制

泌乳乳腺中的上皮细胞不断地分泌乳汁,这些上皮细胞排列在血管周围,从血液中吸收各种制备乳汁所需的前体物质,从而合成乳汁并将其释放到腺泡腔中。当病原体入侵并破坏乳腺屏障时,就会造成组织损伤,并向乳腺系统释放多种不同的化合物。此时,上皮细胞通过限制感染的程度和激活乳腺免疫系统来发挥防御功能。白细胞主要包括正常存在于健康乳房中的中性粒细胞、淋巴细胞和巨噬细胞以及被乳腺免疫系统激活的其他细胞。免疫球蛋白包含抗体、补体蛋白和抗菌肽等。每次挤奶时,乳头括约肌打开,从而使乳腺受到感染的威胁,这促使了白细胞的激活。当乳腺内的常驻巨噬细胞识别到某些有害细菌时,它们就会向其他白细胞和免疫细胞发出信号,招募更多的免疫细胞到感染区域,造成了牛奶中的体细胞数增加。吞噬细胞,即中性粒细胞和巨噬细胞,在感知到外源性细菌感染后,会锁定并吞噬细菌,然后通过释放酶消化细菌成分以杀灭细菌。在炎症期间,中性粒细胞是发挥主要作用的白细胞,也被称为从血液进入乳腺的第一道防线。中性粒细胞能够吞噬并杀死微生物,有效抵御感染。综上所述,感染期间乳汁体细胞的释放见图 1-11。

四、乳汁中白细胞种类及作用

在发生乳腺炎时,正常乳腺区的淋巴细胞比例增加,巨噬细胞和中性粒细胞比例也有所上升。然而,也有报道称巨噬细胞在正常乳汁体细胞中所占比例最高。低体细胞数的牛奶中,中性粒细胞可能是调节乳房动态免疫防御的第一步。因此,中性粒细胞百分比变化与体细胞计数结合可被用于评估牛奶质量。有报道称,初乳第 1 天中性粒细胞百分比最高(36.91%),第 5 天最低(24.90%)。淋巴细胞是低、高体细胞数奶牛中的主要细胞,但其比例随着总体细胞数的增加而下降。一些研究者发现,淋巴细胞在正常乳汁和乳腺炎乳汁中所占的体细胞数比例最低。这些白细胞总数约占"正常"牛奶中体细胞数的 25%,它们共同担负着检测和吞噬病原体并随后启动免疫反应的作用。在感染发生时,驻留的巨噬细胞和乳腺上皮细胞首先合成并释放各种促炎细胞因子(如 TNF-α、IL-1 和 IL-8),这些细胞因子能够诱导中性粒细胞从周围血管中趋化进入乳房组织。淋巴细胞则通过释放细胞因子来协调免疫系统反应。巨噬细胞和中性粒细胞都在吞噬和破坏细菌中发挥作用。这些免疫细胞都从血液中进入乳房,作为一种寻找细菌的监视机制,这也是哺乳动物发生乳腺炎的一个主要因素。

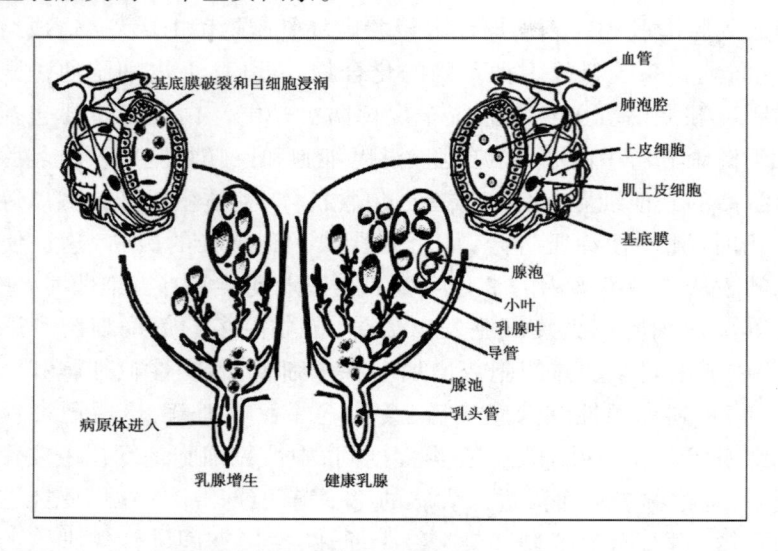

图 1-11 感染期间乳汁体细胞的释放过程

五、影响牛奶体细胞数的因素

牛奶产量、动物健康、管理和环境等均会影响牛奶中体细胞的释放。体细胞数与产奶量之间存在一定的关系,但这种关系受遗传的影响较小。高产奶量的奶牛处于持续产奶的压力下,其免疫力降低,导致牛奶中体细胞数增多。高体细胞数不仅会对产奶量产生负面影响,还会影响牛奶成分和品质。此外,由于产奶量较低,年轻奶牛的体细胞数较低。

奶牛的泌乳期可分为早期、中期和晚期。泌乳早期产奶量最高,随着泌乳的进行而降低。泌乳后不久体细胞数最高,在第 25 至 45 天迅速下降至最低点,然后缓慢上升。首次泌乳的奶牛,乳中的平均体细胞数在泌乳早期较高,为 $(1.10 \sim 1.27) \times 10^5$ 个 /mL,在泌乳中期下降至

$(0.90 \sim 0.99) \times 10^5$ 个/mL，在泌乳后期略有增加，为$(0.99 \sim 1.07) \times 10^5$ 个/mL。总体而言，奶牛的平均体细胞数在泌乳第1个月较高，第2个月下降，此后一直波动至泌乳第300天。哺乳期对乳内感染有明显影响，哺乳期第1个月发生乳内感染的风险最小，第10个月的风险最大，两者相差约6.3倍。乳内感染患病率在泌乳第3个月达到最大（79%），然后在泌乳中期（第6个月）略有下降至60%，在泌乳后期上升至75%左右。高产奶牛乳腺在泌乳中期比泌乳早期和后期表现出更强烈的先天免疫，这体现在乳体细胞数和分离乳白细胞体外免疫应答方面。在瑞典西南部的一项研究中，调查了奶牛首次产犊后第一次试乳时奶牛体细胞数升高（≥20万个/mL），并分析了产犊后第一次试乳（21天）时，18.1%的奶牛的乳汁中体细胞数升高发生率的影响因素。

与多胎奶牛相比，初产奶牛产奶量更少，体细胞数也更低。乳汁体细胞数受胎次影响，初产犊的体细胞数较低。奶牛总体平均体细胞数为51.9×10^3 个/mL，第一胎、第二胎和第三胎细菌阴性体细胞数的最小二乘平均值分别为44.7×10^3 个/mL、50.9×10^3 个/mL和53.0×10^3 个/mL。然而，研究发现第二次泌乳细菌阴性奶牛与第一次泌乳相比较具有显著差异性，而第三次泌乳奶牛与第一次泌乳相比较无显著性。另一项研究发现，第一胎奶牛中未感染的奶牛在泌乳初期体细胞数最高。研究还发现，在整个哺乳期，初产奶牛的乳腺免疫力始终高于多产奶牛。乳体细胞对病原体的反应可能随着年龄的增长而增加，这使乳腺更容易受到新的感染，多胎次奶牛乳房组织的感染时间更长，损伤也更多。

环境因素对奶牛乳汁体细胞数有显著影响。极端温度不仅会给动物带来压力，而且会影响饲料的摄入量。高湿度和质量差的饲料不但会导致微量元素缺乏，也会造成感染性细菌的大量滋生，同时降低免疫力。在产犊模式非季节性的国家，体细胞数在春季和夏季最高，这可能与高温和高湿会增加感染风险有关。产犊前后体细胞数最高的季节是冬季，而最低的体细胞数出现在产犊后不久。与非高产奶牛相比，高产奶牛在湿热季节的体细胞数水平较高，这表明在湿热季节，奶牛乳房承受的压力更大。牛奶中酪蛋白（γ-CN除外）在夏季较低，在冬季较高。而IgG和血清白蛋白含量夏季高于冬季和春季。季节对牛奶体细胞数也有轻微的影响，夏季的体细胞数大于冬季和春季。牛奶凝固性在夏季变差，乳汁体细胞数和中性粒细胞/巨噬细胞比值在冬季居中，夏季最高。在夏季，奶牛乳汁体细胞数和中性粒细胞/巨噬细胞比值均表现出昼夜节律。

随着奶牛产奶能力的提高，所有大的集约化养殖场都引进了挤奶机。机器挤奶需要适当的清洁，顺利运作，并根据制造商的规格进行维护。研究发现，牛奶体细胞数与挤奶方式之间有关系，手工挤奶的乳体细胞数高于机器挤奶。挤奶后，乳头药浴可以减少之后牛奶中的体细胞数。现代自动机器内置了体细胞数在线计数器，一旦牛奶通过，它就会显示乳体细胞数。研究发现，在安装自动挤奶系统12个月后，体细胞数明显升高，随着时间的推移而减少，甚至在安装36个月后体细胞数显著降低。

不同品种奶牛的体细胞数存在差异，如图1-12所示。高产牛品种，如瑞士褐牛和荷斯坦牛，在牛奶中有较高的体细胞存在。212个奶牛群的平均体细胞数为（241 000±83 000）个/mL。品种类型也会影响乳房的形状，乳房附着良好的奶牛患乳腺炎的概率低于乳房下垂型奶牛。乳体细胞数在较短且乳头管直径较大的乳头中较高。

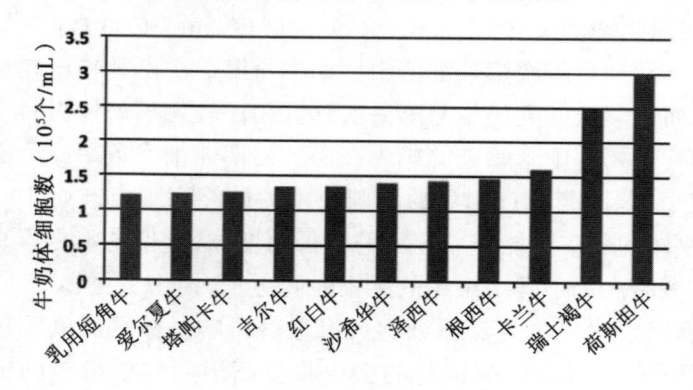

图 1-12 不同品种健康奶牛的乳体细胞计数

体细胞数在患有临床型奶牛乳腺炎的奶牛中最高（7.50×10^5个/mL），其次是患有亚临床奶牛乳腺炎的奶牛（4.60×10^5个/mL），然后是产犊当天收集的初乳（4.00×10^5个/mL），最后是怀孕奶牛（2.00×10^5个/mL）。与体细胞数不同，患有临床型奶牛乳腺炎的奶牛、犊牛和患有亚临床型奶牛乳腺炎的奶牛的乳中性粒细胞百分比最高。健康奶牛的乳巨噬细胞比例较高（65.53%），而普通奶牛和产犊奶牛的乳巨噬细胞比例均显著降低（16.59%）。与中性粒细胞和巨噬细胞相比，乳淋巴细胞在各组间的波动较小。

在哺乳动物的乳腺组织中可以发现两种类型的细菌或病原体：传染性和环境性。传染性病原体在牛与牛之间传播，环境病原体则存在于牛群周围的环境中，如垫料、粪便和土壤等。金黄色葡萄球菌、无乳链球菌、停乳链球菌被归为传染性病原体，它们能有效地适应乳腺环境，并在挤奶过程中在奶牛之间传播。结核菌、肠球菌、化脓性隐球菌、凝固酶阴性葡萄球菌、大肠菌群等病原体被归类为环境致病菌，它们被认为是乳腺的机会性致病菌。在感染金黄色葡萄球菌的水牛乳汁中体细胞数最高，其次是大肠杆菌和无乳链球菌感染的乳汁。然而，在患有乳腺炎的奶牛中，与其他病原体相比，无乳链球菌是导致乳汁中体细胞数更高的原因。牛乳腺中性粒细胞比例在无乳链球菌感染时最高，其次是大肠杆菌和金黄色葡萄球菌感染。此外，乳腺炎的严重程度不影响血液细胞计数，但它影响正常泌乳期的乳汁体细胞数。奶牛在泌乳期有40天的干乳期，这段时间可以让乳腺得到恢复，为下一次泌乳做好准备。干乳期是哺乳周期中最关键的时期之一，在此期间需要采取预防措施，因为乳汁会积聚在乳房内，为细菌提供良好的生长环境。干乳期治疗可以有效减少感染的发生，从而减少下一次泌乳过程中的乳汁体细胞数。

六、高体细胞数对产奶量和乳成分的影响

高体细胞数会影响牛奶质量和成分。高体细胞数通常由乳腺炎症引起，它会像改变血液一样改变牛奶的成分。这是由于血液乳腺屏障的渗透性增加，导致更多的离子、蛋白质和炎症细胞进入牛奶。牛奶体细胞数的增加与产奶量的减少有关。产奶量的减少可以在确诊临床型奶牛乳腺炎前一周左右观察到，这可能是由于在临床型奶牛乳腺炎发病前，乳腺炎已处于亚临床状态。此外，如果感染发生在哺乳早期，且在动物达到产奶量高峰之前，乳腺炎的影响会更严重。与没有患乳腺炎的奶牛相比，患有乳腺炎的奶牛产奶量会减少。奶牛一旦被感染，产奶量便不会恢复到患有乳腺炎前的水平。乳体细胞数升高导致产奶量下降，主要是由于产乳上皮细胞受到物理损伤，导致乳腺的合成和分泌能力显著降低。炎症介质也会改变激素的浓度，无论是刺激还是抑制，乳汁前体的可用性减少都会造成乳腺炎期间乳汁产量减少。此外，更多的能量被转移到免疫系统而不是产奶过程，且奶牛也可能会因为疼痛和运动减少而吃得更少。

牛奶体细胞数对产奶量、乳蛋白和乳糖有显著影响，但对乳脂成分无显著影响。在受感染奶牛的牛奶中，观察到蛋白质含量增加，但乳糖百分比下降。乳糖占牛奶渗透压的 50%，其水平降低会导致牛奶产量急剧下降，更多的离子会从血液转移到牛奶中以维持渗透平衡。健康奶牛的牛奶含有大约 80% 的酪蛋白和 20% 的乳清蛋白。报告表明，乳腺炎奶牛的蛋白质谱发生了变化，乳清蛋白水平有所上升，而 α- 和 β- 酪蛋白减少。这些变化可能是由于哺乳相关基因对感染的反应受到调节，以及当乳汁体细胞数在患有亚临床型乳腺炎和临床型乳腺炎的奶牛牛奶中略有增加时，可观察到的乳蛋白发生诱导水解。此外，蛋白质的增加是由于微生物毒素破坏了乳腺上皮的完整性和紧密连接，导致血源性蛋白质流入乳汁。牛奶中的酪蛋白溶酶，即纤溶酶增加。纤溶酶是由纤溶酶原衍生而来，纤溶酶原来源于血液，可能由于上皮的破坏而渗漏到乳汁中。纤溶酶将 β- 酪蛋白切割成更小的酪蛋白和多肽片段，然后扩散到牛奶中。这导致凝乳效果差，奶酪产量降低，乳制品口感苦涩。

牛奶中粗蛋白质含量在体细胞数约 70 万个 /mL 时略有升高，在体细胞数超过 100 万个 /mL 时有所下降。酪蛋白百分比降低，可溶性蛋白百分比降低，导致酪蛋白与可溶性蛋白比例显著降低。较高的牛奶体细胞数也导致 β- 酪蛋白、α- 酪蛋白和 κ- 酪蛋白含量降低。与先前的研究结果不同，高容积罐奶体细胞数对总蛋白质和乳糖的百分比没有显著影响。

患有乳腺炎奶牛的牛奶中脂肪浓度下降可能是由于乳腺合成和分泌能力下降所致。游离脂肪酸的增加是由于由白细胞产生的脂肪酶或通过脂蛋白水解的纤溶酶改变了乳脂球膜。然而，一些研究人员报道被金黄色葡萄球菌感染的患有临床型奶牛乳腺炎的奶牛牛奶中乳铁蛋白、蛋白质含量和纤溶蛋白升高，酪蛋白 / 蛋白比、钙和磷降低。亚临床型奶牛乳腺炎降低了乳糖、非脂肪固体和总固体含量，但在感染和未感染的猪舍中，蛋白质和脂肪含量没有显著差异。牛奶中乳糖的减少归因于肺泡上皮细胞的损伤。此外，乳腺炎期间，乳汁中的乳糖会通过增殖的细胞旁通路渗出。钾通过细胞旁途径渗漏，因此其浓度降低。相反，在血液中发现的大量钠和氯化物会渗入牛奶中，并超过正常浓度。牛奶体细胞数对奶粉品质及牛奶的其他工艺性状有不良影响。研究还表明，与此相比，每个细胞的纤溶酶原激活剂的活性，中等体细胞数组是低体细胞数组的 4 倍。体细胞通过释放内源性酶对奶酪的工艺特性和质量产生积极影响。

七、牛奶体细胞数的测定方法

显微镜体细胞计数法是一种在野外条件下,经培训的技术人员/兽医可以即刻使用的简便、廉价的方法。在该方法中,收集新鲜牛奶,将 5～10 μL 牛奶均匀涂在显微镜载玻片的 1 cm^2（1cm×1cm）区域上,并在水平位置自然干燥。接着,用 96% 乙醇固定 3 分钟,再次风干,用二甲苯脱脂 10 分钟。之后用 60% 乙醇漂洗,再次风干,然后用亚甲基蓝染色 15 分钟。随后用水漂洗并再次风干载玻片,此时载玻片可用于研究牛奶样品中存在的各种细胞类型,如淋巴细胞、中性粒细胞和巨噬细胞。

许多公司已经开发出了便携式乳扫描仪来估计牛奶体细胞数。全自动体细胞计数仪则使用一种非常敏感的荧光染料,使细胞计数更加准确、可靠和快速。为了用全自动体细胞计数仪计算体细胞,牛奶样品与含有荧光染料（如 Sofia Green）的染色试剂混合。然后,将 12 μL 染色样品置于一次性芯片的测量室中,在几秒钟到 2 分钟的时间内即可完成分析。体细胞计数系统自动对焦在芯片上,并通过灵敏的相机捕获染色细胞。数字图像的分析算法会确定荧光细胞的数量和尺寸,并据此计算它们的浓度。结果会自动显示在屏幕上并通过打印机打印出来,还可以保存并生成报告。最近,流式细胞术方法也已被用于同时鉴定牛奶的细胞类型,这比显微镜方法更可靠。为了使牛奶中体细胞计数在世界范围内统一,国际乳业联合会和国际动物记录委员会启动了一项新项目,旨在建立原料奶中体细胞的国际参考系统。快速体细胞计数应用程序已被开发为手机应用,它可以直接在应用程序中拍摄实际牛奶样品的实时图像,从而节省时间,报告可以立即发送到实验室进行分析。

八、如何在农场层面降低牛奶体细胞数

有许多因素和管理措施会影响牛奶体细胞数的释放,并可能导致其水平的降低或增加。多年来,研究人员发现奶牛场采用的各种管理措施与牛群体细胞数量之间存在关联。例如,在挤奶时戴手套、使用自动取奶器、挤奶后乳头药浴、最后挤奶、每隔一年检查一次挤奶系统、在挤奶

后保持奶牛站立等。在养殖过程中,使用自由畜栏系统、提供沙垫、每次产犊后清洁产犊栏、监测干乳期奶牛是否患有乳腺炎、使用干乳期治疗法、补充微量营养素以及经常进行加州乳腺炎测试等措施有助于降低牛奶体细胞数。

  疫苗和褪黑素也被用于降低牛奶体细胞数。MASTIVAC I 疫苗减少了感染金黄色葡萄球菌的动物数量,改善了乳腺的整体健康状况,提高了产奶量和质量,这种治疗也有助于消除已经感染金黄色葡萄球菌的奶牛的感染,并减少其牛奶中的体细胞数。然而,多价疫苗对金黄色葡萄球菌感染奶牛的体细胞数量没有显著影响。褪黑素已被发现可以改善牛奶质量,增强奶牛的免疫力。研究发现,接受褪黑激素治疗的奶牛体细胞数和血清皮质醇水平较低,这可能是由于这些奶牛的免疫状态得到了改善。

  在奶牛的生产周期中饲喂特定的营养物质也会影响奶牛的免疫防御能力。除了提供能量和蛋白质外,还应定期给奶牛补充各种微量元素、矿物质元素和维生素。抗氧化剂能够保护奶牛身体免受自由基的伤害,并已用于人类和动物,以防止或延缓细胞损伤。给高体细胞数奶牛补充抗氧化维生素,如维生素 A、维生素 C、维生素 E 和 β-胡萝卜素,以及抗氧化矿物质,如硒、锌和铜,可降低体细胞数,使牛奶成分正常化,同时有助于乳腺炎的早期恢复。与未添加无机硒的奶牛相比,添加无机硒可降低金黄色葡萄球菌感染奶牛的乳内感染。在低硒日粮中添加 0.2% 硒也能降低泌乳奶牛的体细胞数。在维生素 A 加 β-胡萝卜素处理降低了奶牛体细胞数水平,改善了乳房健康,增强了免疫力。在过渡期内补充维生素 E、铜和锌可以有效降低奶牛的体细胞数。补充维生素 E 可以从两个方面影响牛奶质量:降低牛奶中体细胞数和降低主要蛋白水解酶即纤溶酶的活性。目前,对于维持牛奶质量所需的维生素 E 量(体细胞数或氧化水平为衡量标准)尚无定论。最后,尽管有大量研究描述了各种抗氧化剂对牛奶质量及其成分的有益作用,但仍有许多研究报告称,膳食抗氧化剂对牛奶中的乳蛋白、乳糖、脂肪、总固体和非脂肪固体没有影响。

参考文献

[1] PETZER I M, KARZIS J, DONKIN E F, et al. Somatic cell count thresholds in composite and quarter milk samples as indicator of bovine intramammary infection status[J].The Onderstepoort Journal of Veterinary Research,2017,84（1）: e1-e10.

[2] SHARMA N, SINGH N K, BHADWAL M S.Relationship of somatic cell count and mastitis: An overview[J].Asian-Australasian Journal of Animal Sciences,2011, 24（3）: 429-438.

[3] ALHUSSIEN M N, DANG A K.Integrated effect of seasons and lactation stages on the plasma inflammatory cytokines, function and receptor expression of milk neutrophils in Sahiwal（Bos Indicus）cows[J].Veterinary Immunology and Immunopathology, 2017,191: 14-21.

[4] RAINARD P, RIOLLET C.Innate immunity of the bovine mammary gland[J].Veterinary Research,2006,37（3）: 369-400.

[5] BOUTINAUD M, JAMMES H.Potential uses of milk epithelial cells: A review[J].Reproduction, Nutrition, Development,2002,42（2）: 133-147.

[6] DANG A, KAPILA S, SINGH C, et al.Milk differential cell counts and compositional changes in cows during different physiological stages[J].Milchwissenschaft-milk Science International,2008,63: 239-242.

[7] ALNAKIP M E, QUINTELA-BALUJA M, BÖHME K, et al.The immunology of mammary gland of dairy ruminants between healthy and inflammatory conditions[J].Journal of Veterinary Medicine, 2014,2014: 659801.

[8] SWAIN D K, KUSHWAH M S, KAUR M, et al.Formation of NET, phagocytic activity, surface architecture, apoptosis and expression of toll-like receptors 2 and 4（TLR2 and TLR4）in neutrophils of mastitic cows[J].Veterinary Research Communications,2014,38（3）: 209-219.

[9] SWAIN D K, KUSHWAH M S, KAUR M, et al. Neutrophil dynamics in the blood and milk of crossbred cows naturally infected with Staphylococcus aureus[J].Veterinary World,2015,8(3): 336-345.

[10] DOSOGNE H, VANGROENWEGHE F, MEHRZAD J, et al.Differential leukocyte count method for bovine low somatic cell count milk[J].Journal of Dairy Science,2003,86(3): 828-834.

[11] BURVENICH C, VAN MERRIS V, MEHRZAD J, et al. Severity of E.coli mastitis is mainly determined by cow factors[J]. Veterinary Research,2003,34(5): 521-564.

[12] ALHUSSIEN M, MANJARI P, MOHAMMED S, et al.Incidence of mastitis and activity of milk neutrophils in Tharparkar cows reared under semi-arid conditions[J].Tropical Animal Health and Production,2016,48(6): 1291-1295.

[13] VISHNOI P, DANG A. Changes in blood and milk DLC and its effect on milk composition in murrah buffaloes (Bubalus bubalis) suffering from clinical mastitis[J].Indian Journal of Dairy Science,2007,60: 286-292.

[14] HAM ED H, EL F EKI A, GARGOURI A. Total and differential bulk cow milk somatic cell counts and their relation with antioxidant factors[J].Comptes Rendus Biologies,2008,331(2): 144-151.

[15] ALHUSSIEN M, KAUR M, MANJARI P, et al.A comparative study on the blood and milk cell counts of healthy, subclinical, and clinical mastitis Karan fries cows[J].Veterinary World,2015,8(5): 685-689.

[16] MUKHERJEE J, DANG A K.Immune activity of milk leukocytes during early lactation period in high and low yielding crossbred cows[J].Milchwissenschaft-milk Science International,2011,66: 384-388.

[17] CINAR M, SERBESTER U, CEYHAN A, et al.Effect of somatic cell count on milk yield and composition of first and second lactation dairy cows[J]. Italian Journal of Animal Science,2015,

14（1）：3646.

[18] SHARMA T, DAS P K, GHOSH P R, et al.Association between udder morphology and in vitro activity of milk leukocytes in high yielding crossbred cows[J].Veterinary World,2017,10（3）：342-347.

[19] KENNEDY B W, SETHAR M S, TONG A K W, et al. Environmental factors influencing test-day somatic cell counts in holsteins[J].Journal of Dairy Science,1982,65（2）：275-280.

[20] SINGH M, LUDRI R S.Somatic cell counts in murrah buffaloes (bubalus bubalis)during different stages of lactation, parity and season[J].Asian-Austranlasian Journal of Animal Sciences,2001,14（2）：189-192.

[21] SINGH M, DANG A K. Somatic cell count of milk[M].Karnal, India：Published by National Dairy Research Institute,2002,1-25.

[22] MORONI P, SGOIFO ROSSI C, PISONI G, et al. Relationships between somatic cell count and intramammary infection in buffaloes[J].Journal of Dairy Science,2006,89（3）：998-1003.

[23] MUKHERJE E J, VARSHNEY N, CHAUDHURY M, et al. Immune response of the mammary gland during different stages of lactation cycle in high versus low yielding Karan Fries crossbred cows[J] . Livestock Science,2013,154（1/2/3）：215-223.

[24] SVENSSON C, NYMAN A K, PERSSON WALLER K, et al . Effects of housing, management, and health of dairy heifers on first-lactation udder health in southwest Sweden[J].Journal of Dairy Science,2006,89（6）：1990-1999.

[25] GONÇALVES J L, CUE R I, BOTARO B G, et al. Milk losses associated with somatic cell counts by parity and stage of lactation[J].Journal of Dairy Science,2018,101（5）：4357-4366.

[26] SARAVANAN R, DAS D N, DE S, ET AL.Effect of season and parity on somatic cell count across zebu and crossbred cattle population[J] . Indian Journal of Animal Research,2015,49（3）：383-387.

[27] GENEUROVA V, HANNS O, GABRIEL B, et al. Somatic cell counts of milk in relation to production factors[J].Zivocisna Vyroba,1993, 38: 359-367.

[28] LA EVEN S H, DELUYKER H, SCHUKKEN Y H, et al. Influence of parity and stage of lactation on the somatic cell count in bacteriologically negative dairy cows[J].Journal of Dairy Science, 1997,80（12）: 3219-3226.

[29] SCHEPERS A J, LAM T J G M, SCHUKKEN Y H, et al.Estimation of variance components for somatic cell counts to determine thresholds for uninfected quarters[J].Journal of Dairy Science,1997,80（8）: 1833-1840.

[30] DANG A K, MUKHERJE E J, CHAUDHURY M, et al. In vitro phagocytic activity of blood and milk neutrophils against Saccharomyces cerevisiae in primiparous and multiparous Karan Fries crossbred cows throughout the dry period and lactation cycle[J].The Indian Journal of Animal Science,2014,84（3）: 262-266.

[31] MORSE D, DE LORENZO M A, WILCOX C J, et al.Climatic effects on occurrence of clinical mastitis[J].Journal of Dairy Science,1988,71（3）: 848-853.

[32] CLEMENTS A C A, PFEIFFER D U, HAYES D.Bayesian spatio-temporal modelling of national milk-recording data of seasonal-calving New Zealand dairy herds[J].Preventive Veterinary Medicine,2005,71（3/4）: 183-196.

[33] MUKHERJEE J, DE K, CHAUDHURY M, et al. Seasonal variation in in vitro immune activity of milk leukocytes in elite and non-elite crossbred cows of Indian sub-tropical semi-arid climate[J].Biological Rhythm Research,2015,46（3）: 425-433.

[34] BERNABUCCI U, BASIRICÒ L, MORERA P, et al.Effect of summer season on milk protein fractions in Holstein cows[J]. Journal of Dairy Science,2015, 98（3）: 1815-1827.

[35] BOMBADE K, KAMBOJ A, ALHUSSIEN M N, et al . Diurnal variation of milk somatic and differential leukocyte counts of

Murrah buffaloes as influenced by different milk fractions, seasons and parities[J].Biological Rhythm Research,2018,49（1）：151-163.

[36] DANG A，ANAND S.Effect of milking systems on the milk somatic cell counts and composition[J].Livestock Research for Rural Development,2007, 19：1-9.

[37] CASTRO Á，PEREIRA J M，AMIAMA C，et al.Typologies of dairy farms with automatic milking system in northwest Spain and farmers' satisfaction[J].Italian Journal of Animal Science,2015, 14（2）：3559.

[38] ALHUSSIEN M，MANJARI P，SHEIKH A A，et al. Immunological attributes of blood and milk neutrophils isolated from crossbred cows during different physiological conditions[J].Czech Journal of Animal Science,2016, 61（5）：223-231.

[39] PAAPE M J，WIGGANS G R，BANNERMAN D D，et al.Monitoring goat and sheep milk somatic cell counts[J]. Small Ruminant Research,2007,68（1/2）：114-125.

[40] KOC A，KIZILKAYA K. Some factors influencing milk somatic cell count of Holstein Friesian and Brown Swiss cows under the Mediterranean climatic conditions[J]. Archives Animal Breeding, 2009,52（2）：124-133.

[41] IVKIĆ Z，ŠPEHAR M，BULIĆ V，et al. Estimation of genetic parameters and environmental effects on somatic cell count in Simmental and Holstein breeds[J]. Mljekarstvo,2012,62：143-150.

[42] DANG A K，KAPILA S，TOMAR P，et al. Relationship of blood and milk cell counts with mastitic pathogens in Murrah buffaloes[J]. Italian Journal of Animal Science,2007,6（sup^2）：821-824.

[43] AHLAWAT K，DANG A K，SINGH C. Relationships of teat and udder shape with milk SCC in primiparous and multiparous Sahiwal cows[J]. Indian Journal of Dairy Science,2008,61：152-66.

[44] SHARMA T, KUMAR DAS P, GHOSH P R, et al.Alteration in the in vitro activity of milk leukocytes during different parity in high yielding cross-bred cows[J]. Biological Rhythm Research,2016, 47（4）: 519-527.

[45] ALHUSSIEN M N, DANG A K. Milk somatic cells, factors influencing their release, future prospects, and practical utility in dairy animals: An overview[J]. Veterinary World,2018,11（5）: 562-577.

[46] MUKHERJEE J, CHAUDHURY M, DANG A K. Alterations in the milk yield and composition during different stages of lactation cycle in elite and non-elite Karan-Fries cross-bred cows(Holstein Fresian x Tharparkar)[J]. Biological Rhythm Research,2017,48 （4）: 499-506.

[47] MUKHERJEE J, CHAUDHURY M, DANG A K. Alterations in the relative abundance of Haptoglobin（Hp）transcripts in total milk somatic cells during different stages of lactation cycle in high yielding cross-bred cows[J]. Biological Rhythm Research,2017, 48（4）: 577-581.

[48] LE MARÉCHAL C, THIÉRY R, VAUTOR E, et al.Mastitis impact on technological properties of milk and quality of milk products-a review[J].Dairy Science and Technology,2011,91（3）: 247-282.

[49] SHUSTER D E, HARMON R J, JACKSON J A, et al. Suppression of milk production during endotoxin-induced mastitis[J]. Journal of Dairy Science, 1991,74（11）: 3763-3774.

[50] SAVIĆ N R, MIKULEC D P, RADOVANOVIĆ R S. Somatic cell counts in bulk milk and their importance for milk processing[J]. IOP Conference Series: Earth and Environmental Science,2017, 85: 012085.

[51] PETROVSKI K.Milk composition changes during mastitis[J]. Dairy Vets Newsletter,2006,23: 7-12.

[52] MALEK DOS REIS C B, BARREIRO J R, MESTIERI L, et al. Effect of somatic cell count and mastitis pathogens on milk composition in Gyr cows[J]. BMC Veterinary Research,2013,9: 67.

[53] SERT D, MERCAN E, AYDEMIR S, et al. Effects of milk somatic cell counts on some physicochemical and functional characteristics of skim and whole milk powders[J]. Journal of Dairy Science, 2016,99(7): 5254-5264.

[54] BOBBO T, CIPOLAT-GOTET C, BITTANTE G, et al.The nonlinear effect of somatic cell count on milk composition, coagulation properties, curd firmness modeling, cheese yield, and curd nutrient recovery[J].Journal of Dairy Science,2016,99(7): 5104-5119.

[55] CHARISMIADOU M, KARLA G, THEODOROU G, et al. The effect of health status of the udder on plasminogen activator activity of milk somatic cells in ovine milk[J]. Small Ruminant Research,2015,133: 54-57.

[56] LI N, RICHOUX R, BOUTINAUD M, et al.Role of somatic cells on dairy processes and products: A review[J].Dairy Science and Technology,2014,94(6): 517-538.

[57] SÁNCHEZ-MACÍAS D, MORALES-DELANU EZ A, TORRES A, et al. Effects of addition of somatic cells to caprine milk on cheese quality[J]. International Dairy Journal,2013,29(2): 61-67.

[58] LI N, RICHOUX R, PERRUCHOT M H, et al.Flow cytometry approach to quantify the viability of milk somatic cell counts after various physico-chemical treatments[J].PLoS One,2015,10 (12): e0146071.

[59] SYRIDION D, LAYEK S S, BEHERA K, et al.Effects of parity, season, stage of lactation, and milk yield on milk somatic cell count, pH and electrical conductivity in crossbred cows reared under subtropical climatic conditions[J].Milchwissenschaft-milk Science International,2012,67: 362 – 365.

[60] VISSIO C, BOUMAN M, LARRIESTRA A J. Milking machine and udder health management factors associated with bulk milk somatic cell count in Uruguayan herds[J]. Preventive Veterinary Medicine,2018,150: 110-116.

[61] LANDIN H, MÖRK M J, LARSSON M, et al. Vaccination against Staphylococcus aureus mastitis in two Swedish dairy herds[J]. Acta Veterinaria Scandinavica, 2015, 57: 81.

[62] YANG M H, SHI J M, TIAN J H, et al. Exogenous melatonin reduces somatic cell count of milk in Holstein cows[J]. Scientific Reports, 2017, 7: 43280.

[63] SANTOSHI P, OBEROI P S, ALHUSSIEN M N, et al. Combined effect of trisodium citrate and vitamin E supplementation during the transition period on body weight and other production parameters in Sahiwal cows[J]. Indian Journal of Dairy Science, 2018, 71: 78-83.

[64] DANG A K, PRASAD S, DE K, et al. Effect of suppleme-ntation of vitamin E, copper and zinc on the in vitro phagocytic activity and lymphocyte proliferation index of peripartum Sahiwal (Bos Indicus) cows[J]. Journal of Animal Physiology and Animal Nutrition, 2013, 97 (2): 315-321.

[65] YANG F, LI X. Role of antioxidant vitamins and trace elements in mastitis in dairy cows[J]. Journal of Advanced Veterinary and Animal Research, 2015, 2 (1): 1-9.

[66] SALAMI S A, GUINGUINA A, AGBOOLA J O, et al. Review: in vivo and postmortem effects of feed antioxidants in livestock: A review of the implications on authorization of antioxidant feed additives[J]. Animal, 2016, 10 (8): 1375-1390.

[67] KRUZE J, CEBALLOS A, STRYHN H, et al. Somatic cell count in milk of selenium-supplemented dairy cows after an intramammary challenge with Staphylococcus aureus[J]. Journal of Veterinary Medicine Series A, 2007, 54 (9): 478-483.

[68] POLITIS I. Reevaluation of vitamin E supplementation of dairy cows: Bioavailability, animal health and milk quality[J]. Animal: an International Journal of Animal Bioscience, 2012, 6 (9): 1427-1434.

[69] CASTILLO C, PEREIRA V, ABUELO Á, et al. Effect of supplementation with antioxidants on the quality of bovine milk

and meat production[J]. The Scientific World Journal, 2013, 2013: 616098.

[70] WANG Y M, WANG J H, WANG C, et al. Effect of dietary antioxidant and energy density on performance and anti-oxidative status of transition cows[J]. Asian-Australiasian Journal of Animal Sciences, 2010, 23 (10): 1299-1307.

第二章 大肠杆菌感染致奶牛乳腺炎的致病机制

第一节 大肠杆菌感染诱导免疫应答的机制研究

一、概述

抗菌药物或抗生素是从微生物中提取的天然化合物或人工合成药物（如磺胺类药物和喹诺酮类药物），具有阻止其他或目标微生物生长的能力，因此，它们被广泛用于治疗致病菌引起的动物和人类疾病。在兽医领域，抗菌药物的使用目的与人医不同，特别是在食源性动物上，其主要目的为控制和预防细菌感染、促进生长和用于手术后的治疗。因此，细菌感染的控制和预防是通过抗菌药物的治疗或预防应来实现的，且大量的抗菌剂已被用于农业，特别是作为生长促进剂用于食用动物。仅在美国，每年大约使用抗菌剂 9.45×10^6 kg，其中仅一半提供给病人，其余均用于农业。此外，每年大约有 7×10^6 kg 的抗生素（主要是青霉素和四环素）作为促生长剂用于动物体内，且每年约有 4.5×10^4 kg 的抗菌剂（主要是四环素和链霉素）喷洒在水果和作物上，正是这些抗菌药物的滥用，增加了细菌耐药性的出现和传播，给全世界带来了较大的风险。

奶牛乳腺炎是发生在乳腺组织的炎症，会导致产奶量下降、乳品质受损和繁殖效率下降，造成巨大的经济损失。细菌感染是造成乳腺炎的主要原因，大肠杆菌是临床上牛乳腺炎的常见病因。目前，抗生素仍然是用于防治奶牛乳腺炎和其他细菌性疾病的主要方法。然而，随着抗生素的大量使用，全世界范围内从食源性动物中分离出的产超广谱β-内酰胺酶（ESBL）大肠杆菌流行率逐渐增加。2000年的监测数据表明，大肠杆菌菌株对多种主要的抗生素都存在耐药性，如产生超广谱β-内酰胺酶（包括 *TEM*、*SHV*、*CMY* 和 *CTX-M* 类型）、碳青霉烯酶（包含 *KPC*、*NDM*、*VIM*、*OXA*–48 和 *IMP* 类型）、耐氟喹诺酮类、甲氧苄氨嘧啶氨基糖苷类和质粒介导的黏菌素类等。新的耐药性大肠杆菌的迅速传播会导致大肠杆菌性乳腺炎的爆发。研究发现，约有55%临床分离的大肠杆菌菌株对氟喹诺酮类和甲氧苄啶—磺胺甲恶唑类药物具有耐药性。耐抗生素的大肠杆菌菌株越来越普遍，在伊朗，来自12个奶牛场临床型乳腺炎病例的70株大肠杆菌均对林可霉素耐药，且约60%菌株对新霉素和氨苄西林耐药。在中国4个地区，从奶样中分离出的83株大肠杆菌均对青霉素耐药，且最常检测到的耐药基因为 *blaTEM*，检出率为83.1%。近年来，来源于奶牛乳腺炎的产β-内酰胺酶大肠杆菌菌株的分离率逐渐增加，这可能与持续使用抗生素治疗奶牛乳腺炎有关。Yu等人对从中国宁夏奶牛中分离得到的大肠杆菌菌株进行分析，发现5%的分离株为产ESBL菌株。此外，*CTX-M* 类ESBL是全球最常见的大肠杆菌耐药类型，而 *CTX-M*–15 等位基因是中国及世界范围内ESBL的主要 *CTX-M* 型基因。与此同时，从东亚、印度、英国和坦桑尼亚的牛和其他食源性动物中均检测到 *CTX-M*–15 型产β-内酰胺酶大肠杆菌。虽然此前在牛中也检测到其他类型的 *blaCTX-M* 基因，但 *blaCTX-M*–1 是全球最普遍的ESBL编码基因，可以从动物传播给人，这引起了全世界对兽医和公共卫生的关注。

　　Ali等人从中国内蒙古的奶牛乳腺炎病例中分离出的46株产β-内酰胺酶大肠杆菌中最流行的序列类型（ST）为ST410，约占10%。这四株ST410大肠杆菌菌株均具有多重耐药性，然而，ST410大肠杆菌对奶牛乳腺上皮细胞的体外致病性研究尚未报道。因此，在本研究中，奶牛乳腺上皮细胞分别被感染每株产超广谱β-内酰胺酶大肠杆菌、DH5α和P4大肠杆菌菌株，以探究产超广谱β-内酰胺酶大肠杆菌的致病性。

二、大肠杆菌对奶牛乳腺上皮细胞的黏附率与侵入率

ST410（1）、ST410（2）、ST410（3）、ST410（4）这4株ST410大肠杆菌均能够黏附并侵入奶牛乳腺上皮细胞。在3 h内，它们与DH5α、P4大肠杆菌黏附率见表2-1。其侵入率增长见表2-2。此外，4株ST410和P4大肠杆菌的黏附率和入侵率呈时间效应，且3 h的黏附和入侵能力显著高于0.5 h。大肠杆菌菌株种类和感染时间均能显著影响大肠杆菌黏附与侵入奶牛乳腺上皮细胞（$P < 0.05$）。

表 2-1 大肠杆菌对奶牛乳腺上皮细胞的黏附率（%）

时间（h）	大肠杆菌菌株					
	DH5α	ST410（1）	ST410（2）	ST410（3）	ST410（4）	P4
0.5	0.0^{cx}（0.01）	1.5^{bw}（0.20）	1.2^{bw}（0.44）	1.0^{bw}（0.20）	4.2^{aw}（0.44）	4.3^{aw}（0.38）
1	0.1^{dyx}（0.01）	5.4^{cx}（1.3）	8.9^{bx}（0.60）	7.9^{bx}（0.73）	13.4^{ax}（2.31）	11.5^{ax}（1.89）
2	0.2^{dy}（0.07）	15.0^{cby}（2.70）	13.3^{by}（1.90）	14.7^{aby}（2.78）	17.3^{ay}（1.73）	17.8^{ay}（1.32）
3	1.4^{cz}（0.18）	25.3^{cz}（2.80）	38.5^{abz}（2.60）	35.3^{bz}（2.65）	39.0^{akz}（0.83）	40.7^{az}（3.41）

注：大肠杆菌菌株种类和感染时间均能显著影响大肠杆菌黏附奶牛乳腺上皮细胞。同一行中的 a-d 和同一列中 w-z 的表示同一行或同一列中具有相同字母的组别不具有显著差异性（$P > 0.05$）；同一行或同一列中不具有相同字母的组别具有显著差异性（$P < 0.05$）。

表 2-2 大肠杆菌对奶牛乳腺上皮细胞的侵入率（%）

时间（h）	大肠杆菌菌株					
	DH5α	ST410（1）	ST410（2）	ST410（3）	ST410（4）	P4
0.5	0.00^{cx}（0.000）	0.30^{bw}（0.039）	0.23^{bw}（0.881）	0.21^{bw}（0.470）	0.83^{aw}（0.088）	0.80^{ax}（0.076）
1	0.01^{dx}（0.004）	1.07^{cx}（0.254）	1.82^{bcx}（0.088）	1.57^{cx}（0.941）	2.68^{ax}（0.462）	2.49^{aby}（0.057）
2	0.05^{by}（0.007）	3.000^{ay}（0.937）	2.67^{ay}（0.385）	2.93^{ay}（0.555）	3.47^{ay}（0.347）	3.19^{ay}（0.816）
3	0.09^{bz}（0.015）	5.05^{bz}（0.549）	7.70^{az}（0.519）	7.06^{az}（0.529）	7.80^{az}（0.165）	7.96^{az}（0.799）

注：大肠杆菌菌株种类和感染时间均能显著影响大肠杆菌侵入奶牛乳腺上皮细胞。同一行中的 a-d 和同一列中 w-z 的表示同一行或同一列中具有相同字母的组别不具有显著差异性（$P > 0.05$）；同一行或同一列中不具有相同字母的组别具有显著差异性（$P < 0.05$）。

三、大肠杆菌感染对奶牛乳腺上皮细胞炎症因子 IL-1β、IL-6、IL-10 和 TNF-α 释放的影响

4 株 ST410 大肠杆菌均能够促进 IL-1β（图 2-1 中的 A）、IL-6（图 2-1 中的 B）、IL-10（图 2-1 中的 C）和 TNF-α（图 2-1 中的 D）的释放。感染了 4 株 ST410 和 P4 大肠杆菌的奶牛乳腺上皮细胞上清液中 IL-1β、IL-6、IL-10 和 TNF-α 的含量显著高于空白对照组和 DH5α 组（$P < 0.05$）。此外，大肠杆菌感染组诱发的炎症反应显著强于 ST410（1、2、3）组（$P < 0.05$）。

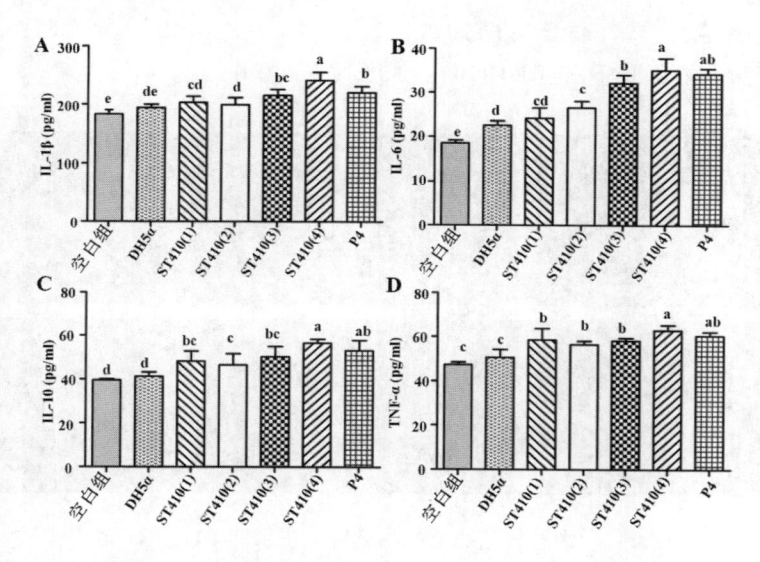

图 2-1　大肠杆菌感染奶牛乳腺上皮细胞上清液中 IL-1β（A）、IL-6（B）、IL-10（C）和 TNF-α（D）的含量

空白组为无感染组，DH5α 为 DH5α 大肠杆菌感染组，ST410（1）为 ST410（1）大肠杆菌感染组，ST410（2）为 ST410（2）大肠杆菌感染组，ST410（3）为 ST410（3）大肠杆菌感染组，ST410（4）为 ST410（4）大肠杆菌感染组 P4 为 P4 大肠杆菌感染组。

四、大肠杆菌感染对奶牛乳腺上皮细胞 ROS 产生、MDA 含量及 GSH-px 和 SOD 活性的影响

4 株 ST410 大肠杆菌感染能够显著地增加奶牛乳腺上皮细胞内 ROS 的产生（图 2-2 和图 2-3 中的 A）和 MDA 含量（图 2-3 中的 B），且显著地降低了 GSH-px（图 2-3 中的 C）和 SOD 活性（图 2-3 中的 D）。ST410（2、3、4）和 P4 组细胞内 ROS 产生和 MDA 含量显著高于空白组和 DH5α 组（$P < 0.05$）；且 GSH-px 和 SOD 活性显著低于空白组和 DH5α 组（$P < 0.05$）。此外，在大肠杆菌感染组，细胞内 ROS 产生和 MDA 含量显著高于 ST410（1、2、3）组（$P < 0.05$）；而 GSH-px 和 SOD 活性显著低于 ST410（1、2、3）组（$P < 0.05$）。

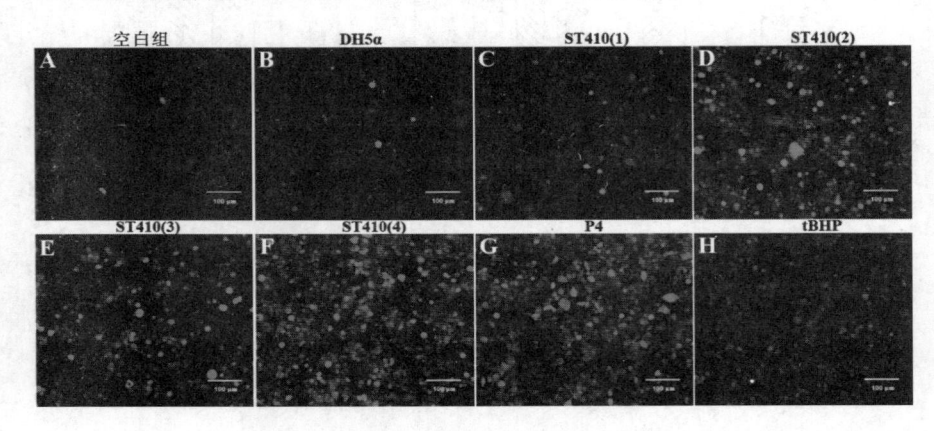

图 2-2 大肠杆菌感染奶牛乳腺上皮细胞 ROS 的产生

空白组为无感染组，DH5α 为 DH5α 大肠杆菌感染组，ST410（1）为 ST410（1）大肠杆菌感染组，ST410（2）为 ST410（2）大肠杆菌感染组，ST410（3）为 ST410（3）大肠杆菌感染组，ST410（4）为 ST410（4）大肠杆菌感染组，P4 为 P4 大肠杆菌感染组，tBHP 为 5 mM tBHP 处理组。

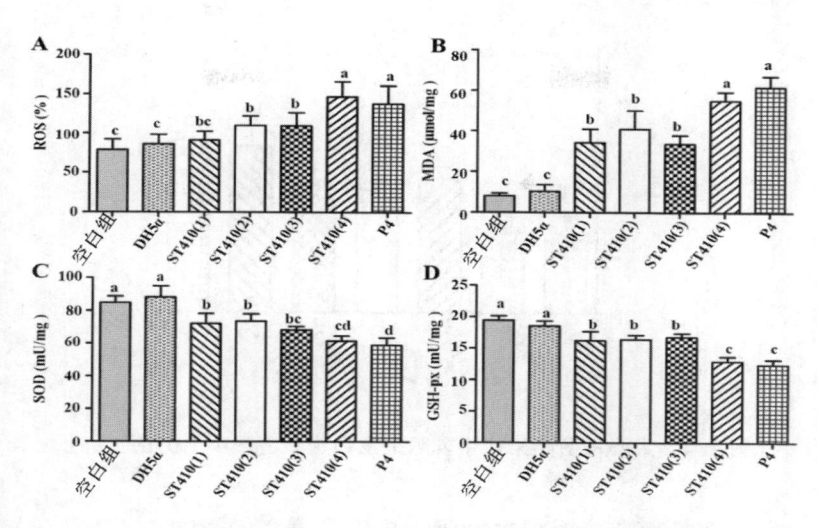

图2-3 大肠杆菌感染奶牛乳腺上皮细胞ROS（A）、MDA含量（B）、SOD（C）和GSH-px（D）的活性

空白组为无感染组，DH5α为DH5α大肠杆菌感染组，ST410（1）为ST410（1）大肠杆菌感染组，ST410（2）为ST410（2）大肠杆菌感染组，ST410（3）为ST410（3）大肠杆菌感染组，ST410（4）为ST410（4）大肠杆菌感染组，P4为P4大肠杆菌感染组，tBHP为5 mM tBHP处理组。

五、大肠杆菌感染对奶牛乳腺上皮细胞上清液中乳酸脱氢酶（LDH）活性的影响

4株ST410大肠杆菌感染能够显著地增加奶牛乳腺上皮细胞上清液中LDH活性（图2-4）。ST410（1、2、3、4）和P4组LDH活性显著高于空白组和DH5α组（$P < 0.05$）。此外，在大肠杆菌感染组LDH活性显著高于ST410（1、2、3）组（$P < 0.05$）。

空白组为无感染组，DH5α为DH5α大肠杆菌感染组，ST410（1）为ST410（1）大肠杆菌感染组，ST410（2）为ST410（2）大肠杆菌感染组，ST410（3）为ST410（3）大肠杆菌感染组，ST410（4）为ST410（4）大肠杆菌感染组，P4为P4大肠杆菌感染组。

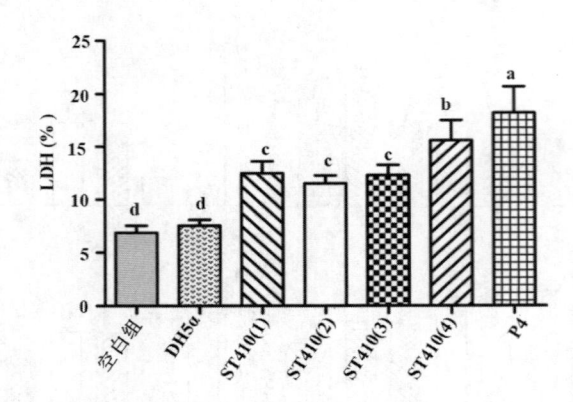

图 2-4　大肠杆菌感染奶牛乳腺上皮细胞上清液中 LDH 的活性

六、大肠杆菌感染对奶牛乳腺上皮细胞凋亡的影响

4 株 ST410 大肠杆菌感染能够显著地增加奶牛乳腺上皮细胞凋亡率（图 2-5）。ST410（1、2、3、4）和 P4 组奶牛乳腺上皮细胞凋亡率显著高于空白组和 DH5α 组（$P < 0.05$）。此外，大肠杆菌感染组奶牛乳腺上皮细胞凋亡率显著高于 ST410（1、2、3）组（$P < 0.05$）。

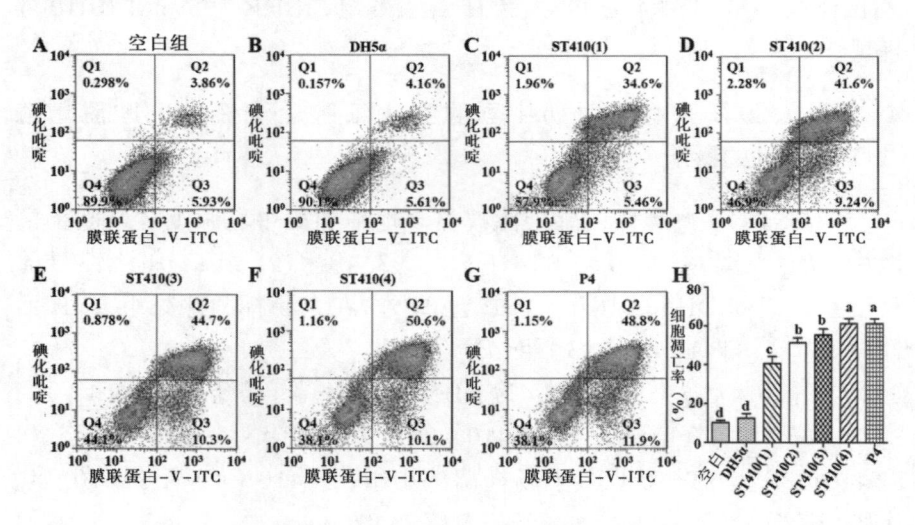

图 2-5　大肠杆菌感染奶牛乳腺上皮细胞的凋亡率

空白组为无感染组，DH5α 为 DH5α 大肠杆菌感染组，ST410（1）为 ST410（1）大肠杆菌感染组，ST410（2）为 ST410（2）大肠杆菌

感染组，ST410（3）为 ST410（3）大肠杆菌感染组，ST410（4）为 ST410（4）大肠杆菌感染组，P4 为 P4 大肠杆菌感染组。

七、讨论

P4（O32：H37）菌株是从英国临床乳腺炎中分离出来的，现在已成为大肠杆菌性乳腺炎研究的典型菌株；而 DH5α 菌株作为非致病性大肠杆菌，只能引起非常轻微的免疫应答，这与它在小鼠、家禽或绵羊等模型中不会引起细菌性疾病的特点保持一致。因此，在本试验中，P4 和 DH5α 分别作为阳性和阴性对照菌株，被用来确定 4 株临床分离的产超广谱 β-内酰胺酶 ST410 大肠杆菌菌株具有致病性。在本研究中，4 株产超广谱 β-内酰胺酶大肠杆菌菌株（ST410）均可以在 0.5 h 内迅速黏附并侵入奶牛乳腺上皮细胞，这与诱导产生急性乳腺感染的大肠杆菌菌株的感染特性相同，且这 4 株 ST410 菌株黏附和侵入奶牛乳腺上皮细胞在 0.5～3 h 之间显著增加。因此，我们推断随着感染时间的延长，黏附和侵入奶牛乳腺上皮细胞的产超广谱 β-内酰胺酶 ST410 大肠杆菌菌株数量增加，呈时间效应。此外，大肠杆菌菌株种类和感染时间均能显著影响大肠杆菌黏附与侵入奶牛乳腺上皮细胞，这表明产超广谱 β-内酰胺酶 ST410 大肠杆菌进入奶牛乳头导管存活、定植，然后造成乳腺组织感染。

大肠杆菌性乳腺炎通常是急性和严重的炎症反应。细胞因子是机体对感染和其他细胞应激反应分泌的重要炎症介质，能够触发免疫细胞活化、分化和募集。IL-1β、IL-6 和 TNF-α 能够促进感染细胞中炎症反应的发生。而 IL-10 能够缓解感染细胞中炎症反应，以调节先天免疫和适应性免疫。此外，LDH 作为细胞质中的糖酵解酶，能够迅速从损伤的细胞中释放到细胞外。故 LDH 能被用于衡量细胞膜的完整性。在本研究中，4 株产超广谱 β-内酰胺酶大肠杆菌菌株（ST410）能够显著增加奶牛乳腺上皮细胞上清液中 LDH 活性和 IL-1β、IL-6、IL-10 和 TNF-α 的含量，该结果与致病性大肠杆菌菌株能够穿透乳腺导管、破坏奶牛乳腺上皮细胞膜，从而刺激白细胞招募引起不同程度的炎症和先天免疫反应相同。研究发现，从感染了大肠杆菌的奶牛收集的牛奶中 IL-6 和 IL-8 含量显著增加，且大肠杆菌感染显著增加了奶牛乳腺上皮细胞和小鼠乳腺上皮细胞中 IL-1β、IL-6、IL-10 和 TNF-α 的含量以及 mRNA 的表达水平。因此,我们推断这 4 株产超广谱 β- 内酰胺酶 ST410 大肠杆菌能够诱导奶牛乳腺上皮细胞发生强烈的炎症反应。此外,IL-10 能通过抑制 IL-1β 和 TNF-α 的作用避免过度炎症反应的发生。因此,本研究中 IL-10 的显著增加能够防止大肠杆菌感染奶牛乳腺上皮细胞发生过度炎症反应。大肠杆菌感染奶牛乳腺后,奶牛乳腺上皮细胞通过释放促炎因子(IL-1β、IL-6 和 TNF-α)来抵抗大肠杆菌感染,同时也促进抗炎因子(IL-10)释放以维持乳腺健康。然而,在本研究中,IL-10 含量的显著升高并没有降低 IL-β、IL-6 和 TNF-α 的释放,这可能是由于 IL-10 不能够完全抑制奶牛乳腺上皮细胞产生的促炎因子(IL-β、IL-6 和 TNF-α)。

  研究发现,大肠杆菌感染奶牛乳腺组织后,能够引发免疫反应并增加 ROS 的产生。当细胞内 ROS 产生量超过 GSH-px 和 SOD 提供的抗氧化缓冲能力时,便会发生脂质过氧化、细胞器降解、脱氧核糖核酸损伤,最终导致细胞死亡。MDA 作为一种氧化终产物,能够反映氧化应激的水平。在本研究中,4 株产超广谱 β- 内酰胺酶 ST410 大肠杆菌感染显著促进奶牛乳腺上皮细胞内 ROS 和 MDA 的产生,同时显著降低 SOD 和 GSH-px 活性。因此,我们推断 ST410 大肠杆菌能够诱导奶牛乳腺上皮细胞发生氧化应激。另有研究发现,奶牛乳腺炎与过度氧化应激相关,表现为 ROS、MDA 和一氧化氮显著升高,谷胱甘肽显著降低。因此,本研究中产超广谱 β- 内酰胺酶 ST410 大肠杆菌感染,诱导奶牛乳腺上皮细胞发生氧化应激,从而损害了奶牛的乳腺健康,降低了牛奶的产量和质量。

  细胞凋亡是乳腺炎发生、发展过程中重要的生物学过程,当该过程因大肠杆菌感染而发生改变时,泌乳中期的荷斯坦奶牛的乳腺细胞因过度凋亡而导致了奶牛乳腺退化。在本研究中,4 株产超广谱 β- 内酰胺酶 ST410 大肠杆菌感染显著增加了奶牛乳腺上皮细胞凋亡指数,这与 Xiu 等人的研究一致。因此,我们推断持续感染大肠杆菌可诱导奶牛乳腺上皮细胞发生细胞凋亡,进而导致乳腺退化和产奶量下降。

  炎症反应是宿主对抗病原体(细菌和病毒)感染的天然保护机制。致乳腺炎大肠杆菌进入乳头管内,能够刺激白细胞募集并诱导乳腺发生急性炎症反应。研究发现,铜绿假单胞菌引起的慢性肺部炎症可通过释放 ROS 引起多核白细胞发生氧化应激后,ROS 激活了凋亡复合物半胱氨酸天冬氨酸蛋白酶(caspase)启动剂,进而切割下游,诱导细胞凋亡。

因此，我们推断随着感染的发生，产超广谱β-内酰胺酶ST410大肠杆菌在诱导炎症反应过程中黏附和侵入奶牛乳腺上皮细胞，从而促进奶牛乳腺上皮细胞发生氧化应激和凋亡，这些细胞内的各种损伤与报道的临床型奶牛乳腺炎的典型临床症状（如发红、肿胀、发热和疼痛）保持一致。

本研究中4株产超广谱β-内酰胺酶ST410大肠杆菌均是内蒙古的同一菌株进化而来，都具有黏附和侵入能力且能够诱导奶牛乳腺上皮细胞发生炎症、氧化应激、细胞膜损伤以及细胞凋亡，其中大肠杆菌感染的致病性显著高于其他3株产超广谱β-内酰胺酶大肠杆菌。因此，大肠杆菌感染对奶牛乳腺上皮细胞的损伤更大，且能触发奶牛乳腺上皮细胞发生更强的炎症、氧化应激、细胞膜损伤以及细胞凋亡。此外，感染了大肠杆菌感染大肠杆菌菌株的奶牛直至扑杀，乳腺持续性肿胀，产奶量下降，被乳中含有大量凝块。因此，大肠杆菌感染是引起临床型乳腺炎的典型菌株，被用于后续研究。

八、小结

临床分离得到的4株多重耐药产超广谱β-内酰胺酶大肠杆菌黏附并侵入奶牛乳腺上皮细胞，引起奶牛乳腺上皮细胞发生炎症、氧化应激、细胞膜损伤和细胞凋亡，这会诱发奶牛乳腺炎并使产奶量下降。其中，大肠杆菌感染的黏附和侵入能力最强，其感染能够触发奶牛乳腺上皮细胞出现最强的免疫应答和过度的细胞凋亡（图2-6），被用于后续研究。

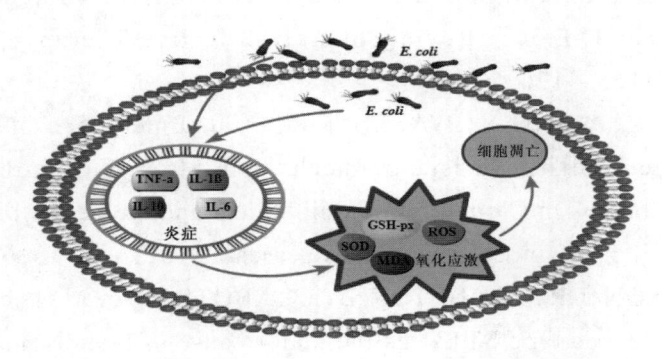

图2-6　大肠杆菌感染奶牛乳腺上皮细胞的机制

参考文献

[1] SCHWARZ S, KEHRENBERG C, WALSH T R. Use of antimicrobial agents in veterinary medicine and food animal production[J]. International Journal of Antimicrobial Agents, 2001, 17（6）: 431-437.

[2] JAMALI H, KRYLOVA K, AÏDER M. Identification and frequency of the associated genes with virulence and antibiotic resistance of Escherichia coli isolated from cow's milk presenting mastitis pathology[J]. Animal Science Joural, 2018, 89（12）: 1701-1706.

[3] DAHMEN S, M ETAYER V, GAY E, et al. Characterization of extended-spectrum beta-lactamase（ESBL）-carrying plasmids and clones of Enterobacteriaceae causing cattle mastitis in France[J]. Veterinary Microbiology, 2013, 162（2/3/4）: 793-799.

[4] GAO J, BARKEMA H W, ZHANG L M, et al. Incidence of clinical mastitis and distribution of pathogens on large Chinese dairy farms[J]. Journal of Dairy Science, 2017, 100（6）: 4797-4806.

[5] UR RAHMAN S, ALI T, ALI I, et al. The growing genetic and functional diversity of extended spectrum beta-lactamases[J]. BioMed Research International, 2018（1）, 9519718.

[6] PITOUT J D D. Extraintestinal pathogenic Escherichia coli: An update on antimicrobial resistance, laboratory diagnosis and treatment[J]. Expert Review of Anti-Infective Therapy, 2012, 10（10）: 1165-1176.

[7] LIU Y Y, WANG Y, WALSH T R, et al. Emergence of plasmid mediated colistin resistance mechanism MCR-1 in animals and human beings in China: A microbiological and molecular biological study[J]. The Lancet Infectious Diseases, 2016, 16（2）: 161-168.

[8] JOHNSON J R, JOHNSTON B, CLABOTS C, et al. Escherichia coli sequence type ST131 as the major cause of serious multidrug-resistant E.coli infections in the United States[J]. Clinical Infectious Diseases, 2010, 51（3）: 286-294.

[9] FAZEL F, JAMSHIDI A, KHORAMIAN B. Phenotypic and

genotypic study on antimicrobial resistance patterns of E.coli isolates from bovine mastitis[J]. Microbial Pathogenesis,2019, 132: 355-361.

[10] YU Z N, WANG J, HO H, et al.Prevalence and antimicrobial-resistance phenotypes and genotypes of Escherichia coli isolated from raw milk samples from mastitis cases in four regions of China[J]. Journal of Global Antimicrobial Resistance,2020,22: 94-101.

[11] KAR D, BANDYOPADHYAY S, BHATTACHARYYA D, et al. Molecular and phylogenetic characterization of multidrug resistant extended spectrum beta-lactamase producing Escherichia coli isolated from poultry and cattle in Odish a, India[J]. Infection, Genetics and Evolution,2015,29: 82-90.

[12] GESER N, STEPHAN R, HÄCHLER H. Occurrence and characteristics of extended-spectrum β-lactamase (ESBL) producing Enterobacteriaceae in food producing animals, minced meat and raw milk[J].BMC Veterinary Research,2012,8 (1): 21.

[13] SAINI V, MCCLURE J T, LEGER D, et al. Antimicrobial resistance profiles of common mastitis pathogens on Canadian dairy farms[J]. Journal of Dairy Science,2012,95 (8): 4319-4332.

[14] YU T, HE T, YAO H, et al.Prevalence of 16S rRNA methylase gene rmtB among escherichia coli isolated from bovine mastitis in Ningxia, China[J].Foodborne Pathogens and Disease,2015,12 (9): 770-777.

[15] LOCATELLI C, CARONTE I, SCACCABAROZZI L, et al. Extendedspectrum β-lactamase production in E.coli strains isolated from clinical bovine mastitis[J].Veterinary Research Communications,2009,33 (1): 141-144.

[16] OHNISHI M, OKATANI A T, HARADA K, et al. Genetic characteristics of CTX-M-type extended-spectrum-β-lactamase (ESBL)-producing enterobacteriaceae involved in mastitis cases on Japanese Dairy farms,2007 to 2011[J]. Journal of Clinical Microbiology,2013,51 (9): 3117-3122.

[17] SU Y C, YU C Y, TSAI Y, et al. Fluoroquinolone-resistant

and extended-spectrum-β-lactamase-producing Escherichia coli from the milk of cows with clinical mastitis in Southern Taiwan[J].Journal of Microbiology, Immunology and Infection, 2016,49（6）:892-901.

[18] UPADHYAY S, HUSSAIN A, MISHRA S, et al. Genetic environment of plasmid mediated CTX-M-15 extended spectrum beta-lactamases from clinical and food borne bacteria in northeastern India[J].PLoS One,2015,10（9）: e0138056.

[19] TIMOFTE D, MACIUCA I E, EVANS N J, et al. Detection and molecular characterization of Escherichia coli CTX-M-15 and Klebsiella pneumoniae SHV-12 β-lactamases from bovine mastitis isolates in the United Kingdom[J].Antimicrobial Agents and Chemotherapy,2014,58（2）: 789-794.

[20] SENI J, FALGENHAUER L, SIMEO N, et al. Multiple ESBL-producing escherichia coli sequence types carrying quinolone and aminoglycoside resistance genes circulating in companion and domestic farm animals in Mwanza, Tanzania, harbor commonly occurring plasmids[J].Frontiers in Microbiology,2016,7: 142.

[21] SEIFFERT S N, HILTY M, PERRETEN V, et al. Extended-spectrum cephalosporin-resistant Gram-negative organisms in livestock: An emerging problem for human health?[J].Drug Resistance Updates, 2013,16（1/2）: 22-45.

[22] HUANG L X, YAO L X, HE Z H, et al. Roxarsone and its metabolites in chicken manure significantly enhance the uptake of As species by vegetables[J].Chemosphere,2014,100: 57-62.

[23] ZHANG H N,ZHOU Y F,GUO S Y,et al. Multidrug resistance found in extended-spectrum beta-lactamase-producing Enterobacteriaceae from rural water reservoirs in Guantao, China[J]. Frontiers in Microbiology,2015,6: 267.

[24] LAUBE H, FRIESE A, VON SALVIATI C, et al. Transmission of ESBL/AmpC-producing Escherichia coli from broiler chicken farms to surrounding areas[J].Veterinary Microbiology,2014,172（3/4）: 519-527.

[25] ALI T, UR RAHMAN S, ZHANG L M, et al. Characteristics and genetic diversity of multi-drug resistant extended-spectrum beta-lactamase（ESBL）-producing Escherichia coli isolated from bovine mastitis[J].Oncotarget,2017,8（52）: 90144-90163.

[26] KEMPF F, SLUGOCKI C, BLUM S E, et al. Genomic comparative study of bovine mastitis escherichia coli[J]. PLoS One,2016,11（1）: e0147954.

[27] CHART H, SMITH H R, LA RAGIONE R M, et al. An investigation into the pathogenic properties of Escherichia coli strains BLR, BL21, DH5α and EQ1[J]. Journal of Applied Microbiology,2000,89（6）: 1048-1058.

[28] BLUM S E, HELLER E D, LEITNER G. Long term effects of Escherichia coli mastitis[J]. Veterinary Journal,2014,201（1）: 72-77.

[29] D ORNE LES E M S, FONSECA M D A M, ABREU J A P, et al. Genetic diversity and antimicrobial resistance in Staphylococcus aureus and coagulase-negative Staphylococcus isolates from bovine mastitis in Minas Gerais, Brazil[J].MicrobiologyOpen, 2019,8（5）: e00736.

[30] DENG S L, YU K, JIANG W Q, et al.Over-expression of Toll-like receptor 2 up-regulates heme oxygenase-1 expression and decreases oxidative injury in dairy goats[J]. Journal of Animal Science and Biotechnology,2017,8（1）: 3.

[31] NEURATH M F, FINOTTO S. IL-6 signaling in autoimmunity, chronic inflammation and inflammation-associated cancer[J]. Cytokine and Growth Factor Reviews,2011,22（2）: 83-89.

[32] MCCOY M K, RUHN K A, BLESCH A, et al. TNF: A key neuroinflammatory mediator of neurotoxicity and neurodegeneration in models of Parkinson's disease[J]. Advances in Experimental Medicine and Biology,2011,691: 539-540.

[33] ZHANG L L, HOU X, SUN L C, et al. Staphylococcus aureus bacteriophage suppresses LPS-induced inflammation in MAC-T bovine mammary epithelial cells[J].Frontiers in Microbiology,

2018,9: 1614.

[34] FRANKE R P, FUHRMANN R, MROWIETZ C, et al. Reduced diagnostic value of lactate dehydrogenase (LDH) in the presence of radiographic contrast media[J]. Clinical Hemorheology and Microcirculation,2010,45（2/3/4）: 123-130.

[35] GILBERT F B, CUNHA P, JENSEN K, et al. Differential response of bovine mammary epithelial cells to Staphylococcus aureus or Escherichia coli agonists of the innate immune system[J]. Veterinary Research,2013,44（1）: 40.

[36] HORWITZ E, KAGAN L, CHAMISHA Y, et al. Novel gastroretentive controlled-release drug delivery system for amoxicillin therapy in veterinary medicine[J]. Journal of Veterinary Pharmacology and Therapeutics,2011,34（5）: 487-493.

[37] GÜNTHER J, ESCH K, POSCHADEL N, et al. Comparative kinetics of Escherichia coli-and Staphylococcus aureus-specific activation of key immune pathways in mammary epithelial cells demonstrates that S.aureus elicits a delayed response dominated by interleukin-6 (IL-6) but not by IL-1A or tumor necrosis factor alpha[J]. Infection and Immunity,2011,79（2）: 695-707.

[38] ZHENG L H, XU Y Y, LU J Y, et al. Variant innate immune responses of mammary epithelial cells to challenge by Staphylococcus aureus, Escherichia coli and the regulating effect of taurine on these bioprocesses[J]. Free Radical Biology and Medicine,2016,96: 166-180.

[39] BIRBEN E, SAHINER U M, SACKESEN C, et al. Oxidative stress and antioxidant defense[J]. World Allergy Organization Journal, 2012,5（1）: 9-19.

[40] SHAHID M, GAO J, ZHOU Y N, et al. Prototheca zopfii isolated from bovine mastitis induced oxidative stress and apoptosis in bovine mammary epithelial cells[J]. Oncotarget,2017,8（19）: 31938-31947.

[41] SOUZA F N, MONTEIRO A M, DOS SANTOS P R, et al. Antioxidant status and biomarkers of oxidative stress in bovine

leukemia virus-infected dairy cows[J].Veterinary Immunology and Immunopathology, 2011,143（1/2）: 162-166.

[42] PILLON BARCELOS R, FREIRE ROYES L F, GONZALEZ-GALLEGO J, et al. Oxidative stress and inflammation: Liver responses and adaptations to acute and regular exercise[J]. Free Radical Research,2017,51（2）: 222-236.

[43] LONG E, CAPUCO A V, WOOD D L, et al. Escherichia coli induces apoptosis and proliferation of mammary cells[J]. Cell Death and Differentiation,2001,8（8）: 808-816.

[44] XIU L, FU Y B, DENG Y, et al. Deep sequencing-based analysis of gene expression in bovine mammary epithelial cells after Staphylococcus aureus, Escherichia coli, and Klebsiella pneumoniae infection[J]. Genetics and Molecular Research,2015, 14（4）: 16948-16965.

[45] HUSSAIN T, TAN B, YIN Y L, et al. Oxidative stress and inflammation: What polyphenols can do for us?[J]. Oxidative Medicine and Cellular Longevity,2016,2016: 7432797.

[46] CIOFU O, RIIS B, PRESSLER T, et al.Occurrence of hypermutable Pseudomonas aeruginosa in cystic fibrosis patients is associated with the oxidative stress caused by chronic lung inflammation[J] . Antimicrobial Agents and Chemotherapy,2005, 49（6）: 2276-2282.

[47] SUN X D, JIA H D, XU Q S, et al. Lycopene alleviates H_2O_2-induced oxidative stress, inflammation and apoptosis in bovine mammary epithelial cells via the NFE2L2 signaling pathway[J]. Food and Function,2019,10（10）: 6276-6285.

[48] ZHUANG C C, HUO W L, LIU G, et al. In vitro immune responses of bovine mammary epithelial cells induced by Escherichia coli, with multidrug resistant extended-spectrum β-lactamase, isolated from mastitic milk[J]. Microbial Pathogenesis,2020,149: 104494.

第二节　大肠杆菌感染诱导乳腺上皮细胞焦亡的机制研究

一、概述

奶牛乳腺炎是由病原微生物引起的乳腺炎症,会造成产奶量和奶品质下降,对动物福利和奶业的营利能力产生不利的影响,甚至威胁公众健康。近几十年来,环境型奶牛乳腺炎的发病率显著增加。大肠杆菌是牛奶中最常见的环境病原体,发病率高,可在奶牛泌乳初期引发乳腺炎。大肠杆菌感染可引起奶牛急性肿胀、疼痛、发热和乳腺疼痛,乳汁呈水样、带血或凝乳状,产奶量减少,甚至导致奶牛死亡。大肠杆菌作为人畜共患病原体,在部分乳制品(如生乳和奶酪制品)中可被检出,对人畜健康构成威胁,并引发公共卫生问题。因此,研究大肠杆菌诱导的牛乳腺炎的潜在机制具有重要意义。

细胞分裂、细胞分化和细胞死亡是维持多细胞生物健康的三个复杂生理过程。真核细胞周期可分为 G1、S、G2 和 M 四个不同的阶段。细胞周期的控制对于调节细胞分化、增殖和死亡等一系列中枢过程至关重要。cdc25c 作为 cdc25 磷酸酶家族的一种亚型,通过 Ser216 磷酸化失活导致其降解,促进细胞周期蛋白依赖性激酶 -1(CDK1)中磷酸化 Tyr15 的活化,抑制细胞周期蛋白 B1/CDK1 复合物的活化,从而抑制细胞增殖,阻滞细胞周期。此外,细胞周期和细胞死亡过程的失调在感染和炎症疾病的发生发展中起着重要作用。凋亡和焦亡作为两种程序性细胞死亡方式,在乳腺炎的发生发展中起着重要作用。虽然细胞凋亡和焦亡是哺乳动物维持体内平衡的生理机制,但一些病原体通过诱导细胞凋亡、焦亡和组织破坏,对宿主对感染的免疫反应产生不利影响。被激活的 NLRP3 通过凋亡相关斑点样蛋白(ASC)激活 caspase-1,最终导致 gasdermin D 裂解形成 N 端片段(GSDMD-N),并释放炎症物质 IL-1β 和 IL-18,导致细胞焦亡。

p53-p21 通路在细胞周期阻滞、细胞凋亡和焦亡中起着至关重要的作用。p53 通过结合 p21 的启动子区激活 p21 基因的转录,进而通过靶向 Bcl-2 蛋白调控细胞凋亡,通过抑制 CDK1 阻滞细胞周期,而焦亡被认为是完全凋亡的结果。然而,大肠杆菌感染是否通过激活 p53-p21 途径介导的奶牛乳腺上皮细胞凋亡和细胞周期阻滞诱导焦亡尚不清楚。因此,本研究的目的是通过建立大肠杆菌感染奶牛乳腺上皮细胞体外实验模型,探讨大肠杆菌感染导致奶牛乳腺上皮细胞焦亡的机制,为预防和管理大肠杆菌乳腺炎提供新的见解,并促进寻找治疗乳腺炎的替代抗生素方法。

二、大肠杆菌感染增加奶牛乳腺上皮细胞 LDH 活性以及 IL-10、IL-1β、IL-18 和 TNF-α 含量

与对照组相比,大肠杆菌感染显著($P<0.01$)提高了 LDH 活性和 IL-10、IL-1β、IL-18 和 TNF-α 浓度(图 2-7)。

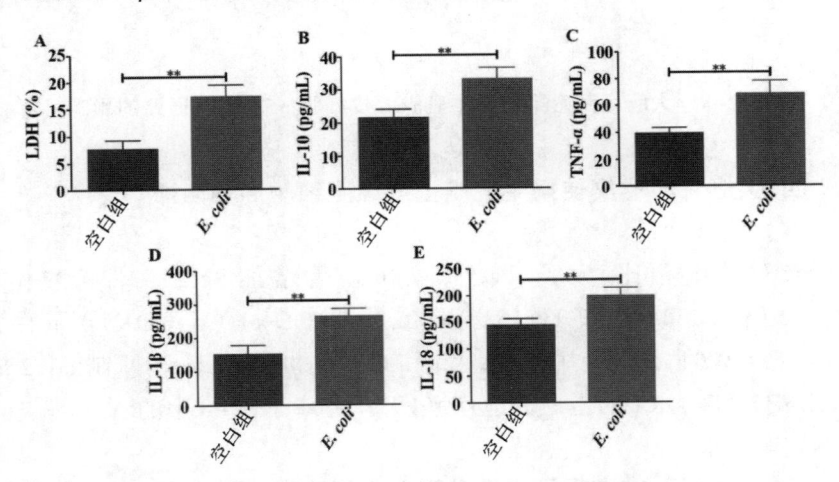

图 2-7　大肠杆菌感染对奶牛乳腺上皮细胞中 LDH 活性(A)和 IL-10(B)、TNF-α(C)、IL-1β(D)和 IL-18(E)含量的影响

三、大肠杆菌感染激活奶牛乳腺上皮细胞内 p53-p21 通路

与对照组相比,大肠杆菌感染的奶牛乳腺上皮细胞 p53 和 p21 蛋白表达水平显著($P<0.01$)增加(图 2-8A,B 和 C)。

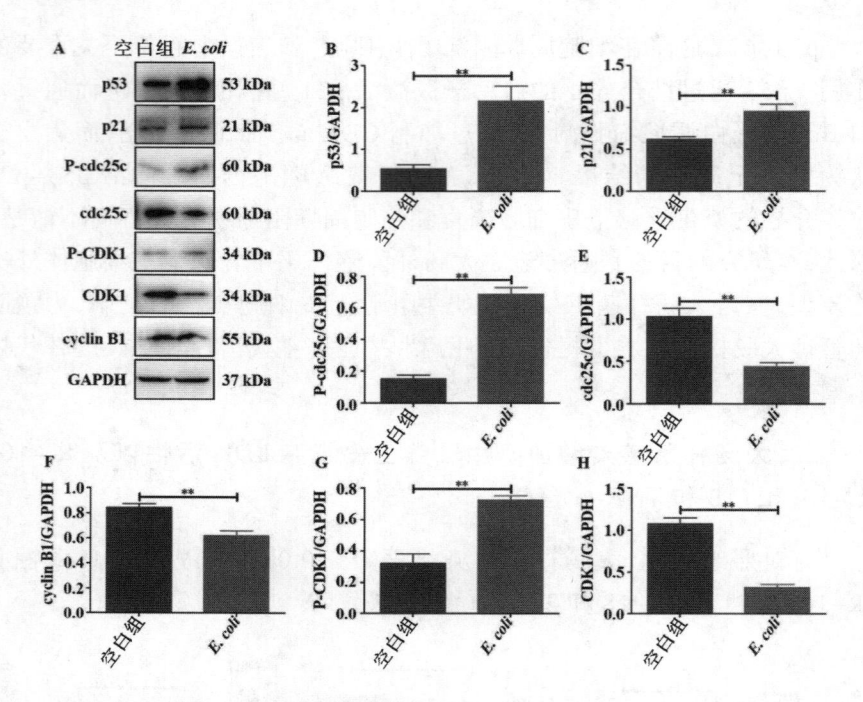

图 2-8 大肠杆菌感染对奶牛乳腺上皮细胞 p53-p21 通路的影响

四、大肠杆菌感染使奶牛乳腺上皮细胞细胞周期阻滞

与对照组相比,大肠杆菌感染可显著增加 Ser216 端磷酸化的 cdc25c(p-cdc25c)和 Tyr15 端磷酸化的 CDK1(p-CDK1)蛋白的表达($P < 0.01$,图 2-8 中的 A、D 和 G),同时可以显著抑制 cdc25c、cyclin B1 和 CDK1 的蛋白表达水平($P < 0.01$,图 2-8 中的 A、E、F、H)。

五、大肠杆菌感染诱导奶牛乳腺上皮细胞发生细胞凋亡

与对照组相比,大肠杆菌感染显著增加了奶牛乳腺上皮细胞的凋亡率($P < 0.01$,图 2-9 中的 A 和 B),以及剪切的 caspase-9、剪切的 caspase-3 和剪切的多聚 ADP 核糖聚合酶(PARP)蛋白表达水平($P < 0.01$,图 2-9 中的 E、G、H、I)。与对照组相比,大肠杆菌感染显著降低了奶牛乳腺上皮细胞的膜电位及 Bcl-2 与 Bcl-2 关联 X 蛋白(Bax)比值($P < 0.01$,图 2-9 中的 C、D、F)。

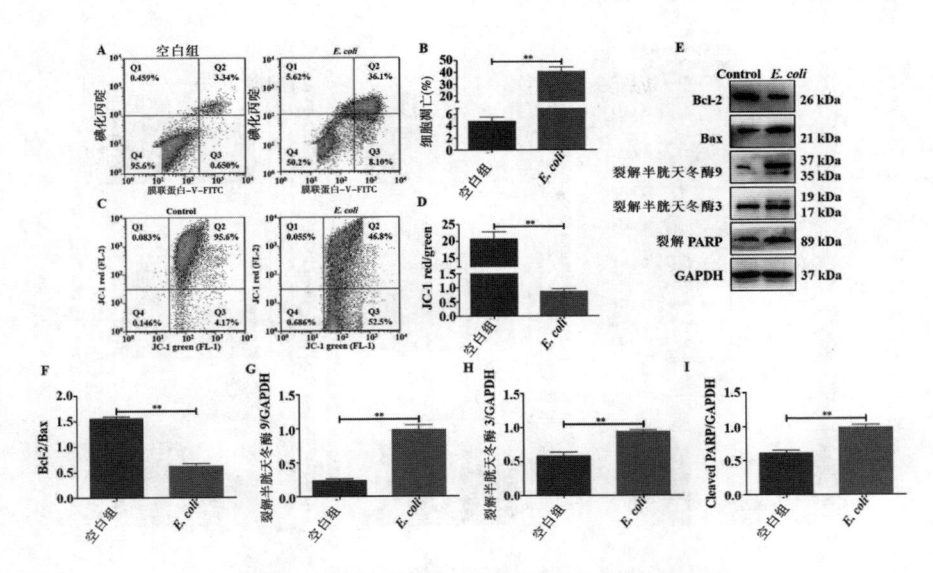

图 2-9　大肠杆菌感染诱导牛乳腺上皮细胞凋亡

六、大肠杆菌感染诱导奶牛乳腺上皮细胞焦亡

与对照组相比，大肠杆菌感染促进了奶牛乳腺上皮细胞中 Ser536 端磷酸化的核因子激活的 B 细胞的 κ-轻链增强子（NF-κB）、NLRP3、ASC、caspase-1、GSDMD-N、IL-1β 和 IL-18 蛋白的表达（$P<0.01$，图 2-10）。

七、p53-p21 通路在细胞周期阻滞、凋亡和焦亡中的作用

为了阐明 p53-p21 通路在大肠杆菌诱导的细胞周期阻滞、凋亡和焦亡中的作用，我们使用了 p53 的特异性抑制剂 2-（2-亚氨基-4,5,6,7-四氢苯并噻唑-3-基）-1-P-苯甲基乙酮氢溴酸盐（Pifithrin-α）抑制 p53-p21 通路。Pifithrin-α 显著抑制了（$P<0.01$）大肠杆菌感染引起的 p53、p21、p-cdc25c、p-CDK1、p-NF-κB、NLRP3、ASC、caspase-1、GSDMD-N、IL-1β、剪切的 caspase-9、剪切的 caspase-3 和剪切的 PARP 蛋白表达水平以及细胞凋亡率的增加（图 2-11 ~ 图 2-13）。Pifithrin-α 显著增加了大肠杆菌感染引起 cdc25c、cyclin B1 和 CDK1 蛋白表达及 Bcl-2 与 Bax 比值的降低。

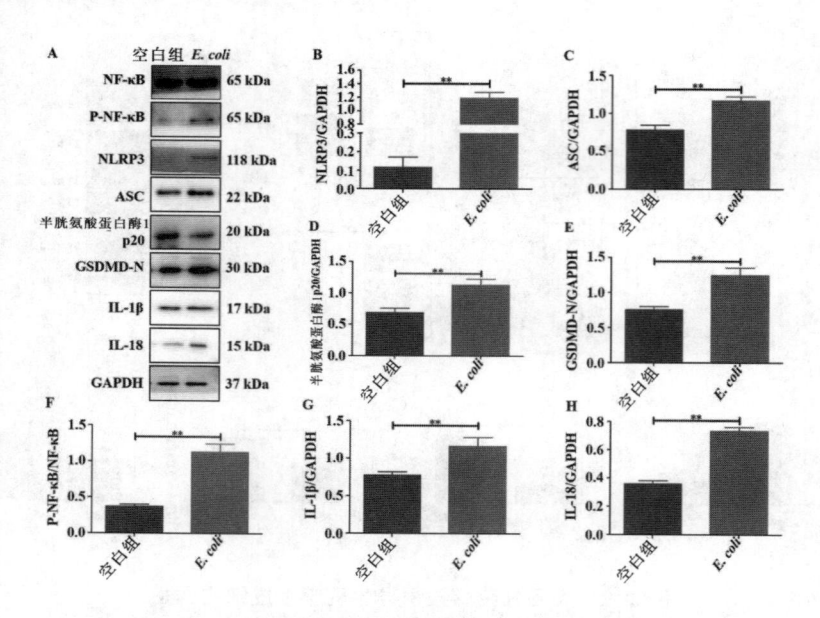

图 2-10　大肠杆菌感染导致奶牛乳腺上皮细胞焦亡

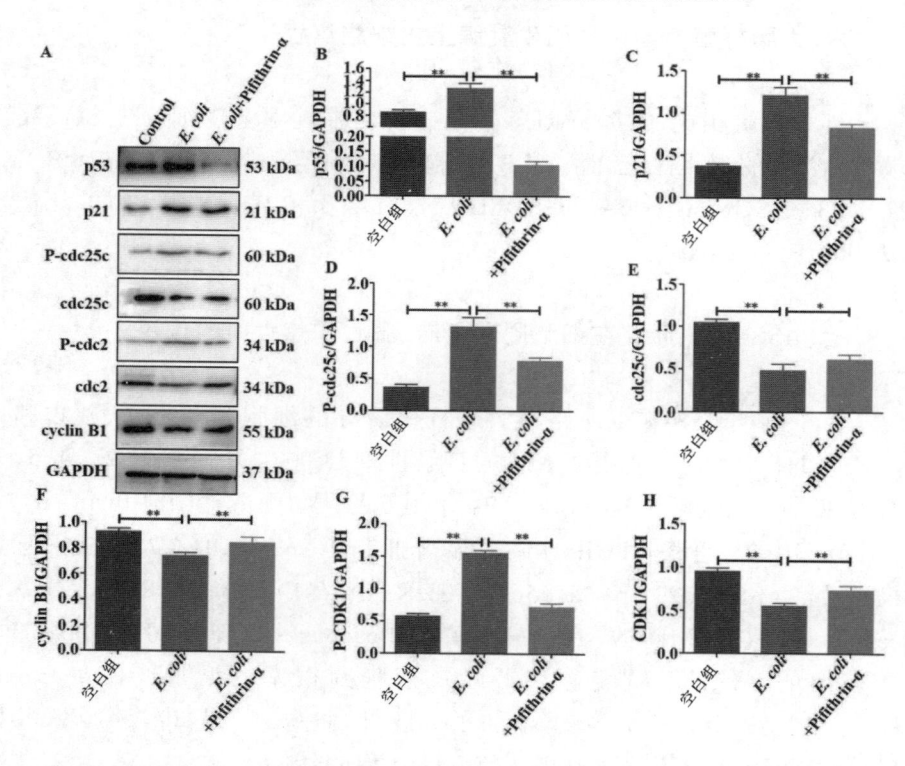

图 2-11　抑制 p53-p21 通路逆转大肠杆菌诱导的奶牛乳腺上皮细胞的细胞周期阻滞

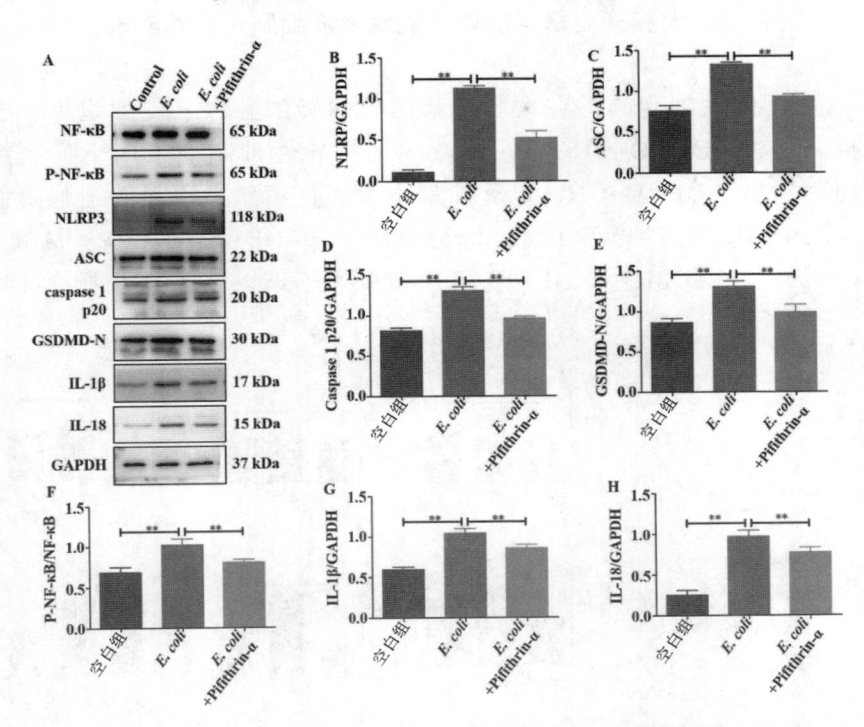

图 2-12　抑制 p53-p21 通路可减轻大肠杆菌诱导的奶牛乳腺上皮细胞焦亡

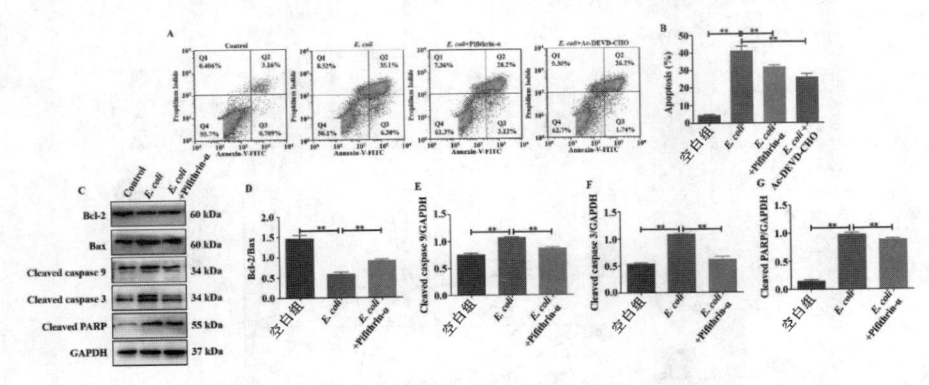

图 2-13　抑制 p53-p21 通路或 caspase-3 可减轻
大肠杆菌诱导的奶牛乳腺上皮细胞凋亡

八、凋亡对大肠杆菌诱导的奶牛乳腺上皮细胞焦亡的影响

为了阐明细胞凋亡在大肠杆菌诱导的奶牛乳腺上皮细胞焦亡中的作用，Ac-DEVD-CHO，caspase-3 的特异性抑制剂被用于抑制细胞凋亡。Ac-DEVD-CHO 显著抑制了大肠杆菌感染引起的细胞凋亡率（$P < 0.01$，图 2-13）、cleaved caspase-3、p-NF-κB、NLRP3、ASC、caspase-1、GSDMD-N、IL-1β 和 IL-18 蛋白表达（$P < 0.01$，图 2-14）的升高。

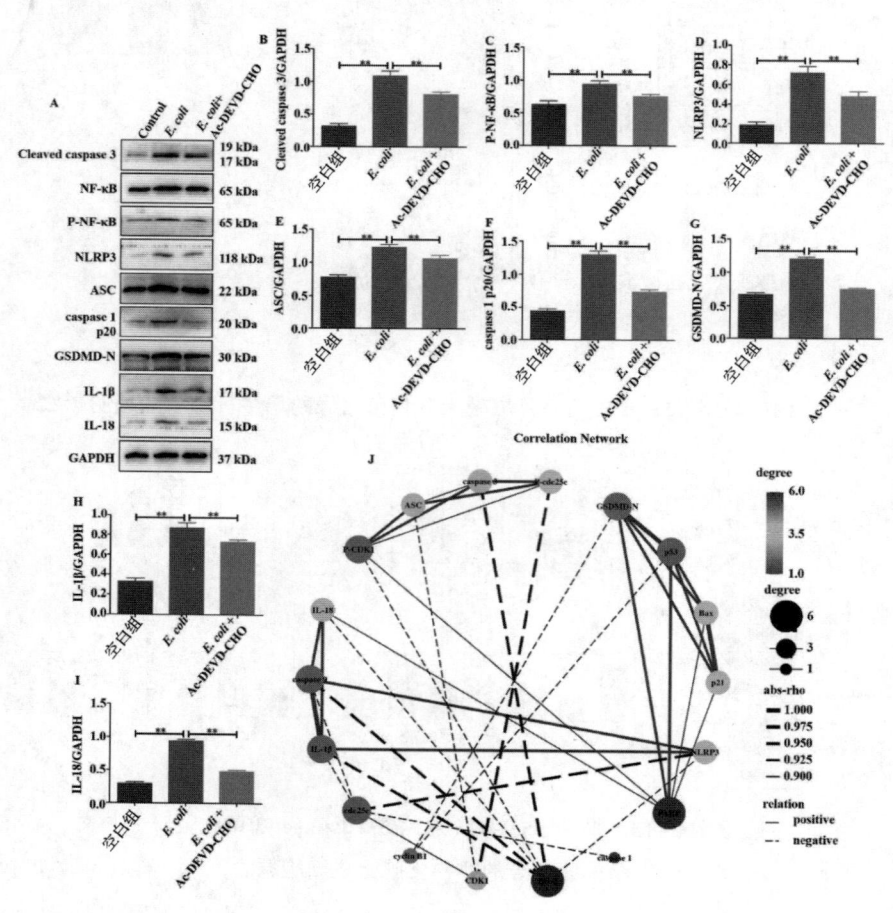

图 2-14　抑制的 caspase-3 可缓解大肠杆菌诱导的奶牛乳腺上皮细胞焦亡

九、讨论

奶牛乳腺上皮细胞通过产生炎性细胞因子在细菌感染的免疫应答中发挥关键作用，因此被认为是研究乳腺炎发病机制的良好体外模型。我们利用奶牛乳腺上皮细胞来探讨大肠杆菌感染诱导的焦亡是否触发了 p53-p21 途径调控的细胞凋亡和细胞周期阻滞。然而，在大肠杆菌感染的奶牛乳腺上皮细胞中，LDH 活性水平升高，表明从患有临床牛乳腺炎的奶牛中分离的大肠杆菌可诱导奶牛乳腺上皮细胞膜损伤并抑制细胞增殖，这与先前的研究结果一致。

细胞周期是细胞分裂过程中发生的一系列有序活动的集合。正常的细胞生命是由调控良好的细胞周期控制的，但在许多疾病的病理生理过程中，细胞周期失控会产生重大影响。各种微生物感染会破坏细胞周期机制，促进病原体在宿主细胞定植。大肠杆菌在细胞周期的 S 期和 G2-M 期捕获细胞。本研究结果表明，大肠杆菌感染显著提高了 p-cdc25c 和 p-CDK1 蛋白水平，抑制了奶牛乳腺上皮细胞中 cdc25c、CDK1 和 cyclin B1 的表达，这与肠致病性大肠杆菌感染肠上皮细胞通过抑制 CDK1 诱导细胞周期阻滞的现象不一致。因此，我们可以得出结论，大肠杆菌感染导致奶牛乳腺上皮细胞周期阻滞。

然而，过度的细胞凋亡会导致乳腺萎缩，我们证明了大肠杆菌在乳腺上皮细胞凋亡中的促凋亡作用。与对照组相比，大肠杆菌感染的细胞中 Bcl-2/Bax 比值和细胞膜电位降低，cleaved caspase-9 和 cleaved caspase-3 蛋白表达升高，凋亡率也显著升高，这证明了大肠杆菌感染的促凋亡作用。因此，大肠杆菌通过 Bcl-2 家族蛋白（Bcl-2 和 Bax）扩大线粒体通透性过渡孔，进而激活 caspase-9 和 caspase-3 蛋白，导致细胞凋亡。

焦亡是一种程序性细胞死亡方式，NLRP3、ASC、caspase-1、GSDMD、IL-18 和 IL-1β 是介导焦亡经典途径的传统标志基因。NLRP3 可被成孔毒素、细胞外三磷酸腺苷、RNA-DNA 杂交分子和病原体激活。激活的 NLRP3 通过激活 caspase-1，将 GSDMD 裂解为两个片段（N 端和 C 端裂解产物，GSDMD-N 和 GSDMD-C）。GSDMD-N 通过释放促炎细胞因子如 IL-1β 和 IL-18 促进焦亡，以应对微生物感染和危险信号。因此，大肠杆菌感染可引起奶牛乳腺上皮细胞发生焦亡。此外，焦亡的

大肠杆菌感染致奶牛乳腺炎的致病机制及防治

发生还伴有严重的炎症反应。因此,本研究中大肠杆菌感染的奶牛乳腺上皮细胞中 IL-1β、TNF-α 和 IL-18 浓度显著升高,提示大肠杆菌感染通过激活奶牛乳腺上皮细胞的焦亡引起炎症反应。此外,IL-10 是一种抗炎细胞因子,可抑制促炎细胞因子如 IL-1β、TNF-α 和 IL-6 的产生,以避免炎症风暴的发生。因此,大肠杆菌感染的奶牛乳腺上皮细胞中 IL-10 浓度显著升高,表明奶牛乳腺上皮细胞通过释放抗炎因子 IL-10 来抑制大肠杆菌感染引起的过度炎症反应(IL-1β 和 TNF-α 表达)。

PARP 是一种多功能蛋白质翻译后修饰酶,通过识别结构损伤的 DNA 片段而被激活。PARP 被认为是 DNA 损伤的受体。因此,本研究中大肠杆菌组 cleaved PARP 水平显著升高,表明大肠杆菌感染通过细胞凋亡和细胞周期阻滞介导的 DNA 断裂激活 PARP。此外,PARP 作为一种转录共调节因子,与 NF-κB 的调控有关,然而,PARP-1 在 NF-κB 功能调节中的作用存在争议,可能是细胞类型和刺激特异性的。在本研究中,我们得出大肠杆菌感染通过激活 PARP 促进 NF-κB 磷酸化的结论。NF-κB p65 是 NLRP3 炎性小体激活启动信号的中心介质,通过诱导 NLRP3 的转录表达起作用。此外,caspase-3 抑制剂 Ac-DEVD-CHO 显著抑制了大肠杆菌感染奶牛乳腺上皮细胞的总凋亡率、p-NF-κB、NLRP3、ASC、caspase-1、GSDMD-N、IL-1β 和 IL-18 的表达,表明 Ac-DEVD-CHO 抑制凋亡后通过降低 NF-κB 来抑制焦亡活化。综上所述,大肠杆菌感染通过促进 NF-κB 诱导细胞凋亡激活、细胞周期阻滞和焦亡,导致牛乳腺炎的发生。

有趣的是,p53 蛋白与 p21 启动子中两个高度保守的 p53 元件结合,在应激刺激下增加 p21 的表达。通过与 cdc25c 通路相互作用激活 p21 阻滞细胞周期,直接靶向 Bcl-2 蛋白引起细胞凋亡。研究结果表明,大肠杆菌感染激活了 p52-p21 途径。此外,p53 的抑制剂 Pifithrin-α 抑制了细胞凋亡和焦亡的激活,缓解了细胞周期阻滞。而肠致病性和肠出血性大肠杆菌产生的周期抑制因子通过 p53 和 p21 蛋白积累阻断宿主细胞周期。大肠杆菌通过激活涉及 p53 的死亡受体通路来增加细菌性前列腺炎的细胞凋亡。因此,我们可以推断,大肠杆菌感染通过激活 p53-p21 途径介导的细胞周期阻滞和凋亡的发生而诱导焦亡。

炎症反应作为宿主抵抗乳腺细菌感染的天然保护机制,通过释放 ROS 引起多形核白细胞氧化应激。然后,ROS 通过启动 caspase 级联反应导致细胞凋亡,而 caspase 级联反应通过介导 GSDMD 的裂解而导致细胞凋亡。综上所述,我们推断随着感染的发生,引起炎症反应,从而促进氧化应激,随后发生细胞凋亡,最终导致奶牛乳腺上皮细胞焦亡,这些变化与体内报道的乳腺炎的典型临床症状(如发红、肿胀、发热、疼痛和产奶量减少)一致。

十、结论

大肠杆菌感染通过激活 p53-p21 途径介导的奶牛乳腺上皮细胞凋亡和细胞周期阻滞来诱导细胞焦亡(图 2-15)。在大肠杆菌诱导的乳腺炎中,通过抑制 p53-p21 途径下调焦亡不会改变奶牛乳腺的功能。本试验能够为大肠杆菌感染引起的牛乳腺炎的预防和治疗提供新思路。

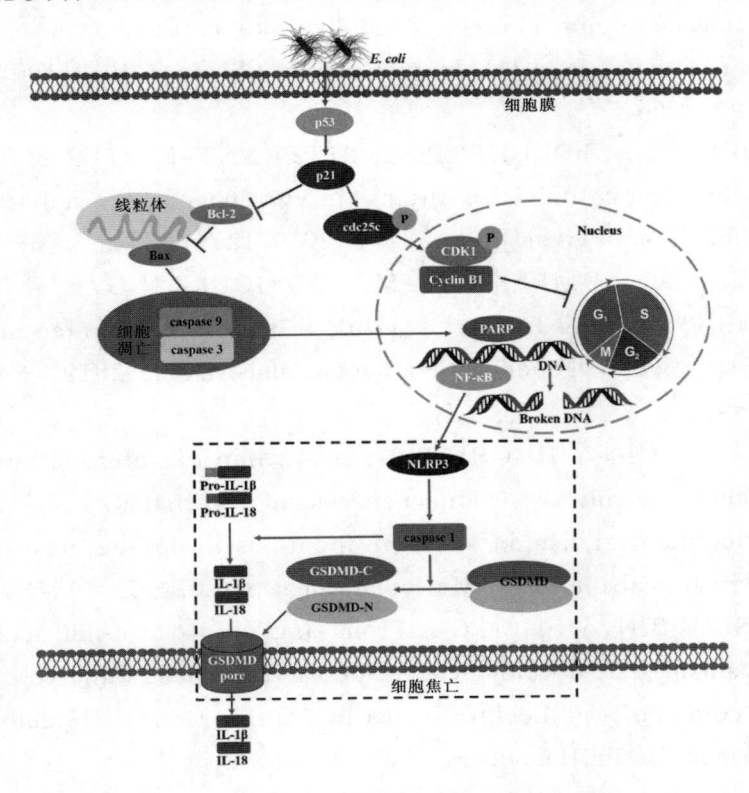

图 2-15 大肠杆菌感染诱导细胞焦亡的信号通路

参考文献

[1] CVETNIC L, SAMARDZIJA M, DUVNJAK S, et al. Multi Locus Sequence Typing and spa Typing of Staphylococcus aureus Isolated from the Milk of Cows with Subclinical Mastitis in Croatia[J]. Microorganisms, 2021, 9（4）: 725.

[2] KLAAS I C, ZADOKS R N. An update on environmental mastitis: Challenging perceptions[J]. Transboundary and Emerging Diseases, 2017, 65: 166-185.

[3] GAO J, BARKEMA H W, ZHANG L M, et al. Kastelic, B. Han, Incidence of clinical mastitis and distribution of pathogens on large Chinese dairy farms[J]. Journal of Dairy Science, 2017, 100（6）: 4797-4806.

[4] ROUSSEL P, PORCHERIE A, RÉPÉRANT-FERTER M, et al. Escherichia coli mastitis strains: in vitro phenotypes and severity of infection in vivo[J]. PLoS ONE, 2017, 12: e0178285.

[5] MURINDA S E, IBEKWE A M, RODRIGUEZ N G, et al. Shiga toxin-Producing Escherichia coli in mastitis: An international perspective[J]. Foodborne Pathogehs and Disease, 2019, 16（4）: 229-243.

[6] LI Y N, ZHU Y H, CHU B X, et al. Map of enteropathogenic Escherichia coli targets mitochondria and triggers DRP-1-mediated mitochondrial fission and cell apoptosis in bovine mastitis[J]. International Journal of Molecular Sciences, 2022, 23（9）: 4907.

[7] VASIEV B N, WEIJER C J. From single cells to a multicellular organism: The development of the social amoeba e dictyostelium discoideum[M]// Lecture Notes in Physics Berlin, Heidelberg: Springer Berlin Heidelberg, 2007: 559-583.

[8] HUANG S Y, PAN J L, KARTHIK S, et al. MEDB-08. Inhibition of different mitotic targets demonstrated distinct DNA damage and cell death response in p53-mutant medulloblastoma[J]. Neuro-Oncology, 2022, 24（Supplement_1）: i105.

[9]SOL A, LIPO E, DE JESÚS-DÍAZ D A, et al. Legionella pneumophila translocated translation inhibitors are required for bacterial-induced host cell cycle arrest[J]. Proceeding of the National Academy of Sciences of the United States of America, 2019,116（8）:3221-3228.

[10]LIU K, ZHENG M Y, LU R, et al. The role of CDC25C in cell cycle regulation and clinical cancer therapy: A systematic review[J]. Cancer Cell International,2020,20: 213.

[11]WIMAN K G, ZHIVOTOVSKY B. Understanding cell cycle and cell death regulation provides novel weapons against human diseases[J]. Journal of Internal Medicine,2017,281（5）:483-495.

[12]SHAHID M, GAO J, ZHOU Y N, et al. Prototheca zopfii isolated from bovine mastitis induced oxidative stress and apoptosis in bovine mammary epithelial cells[J]. Oncotarget,2017,8（19）:31938-31947.

[13]LIU N, WANG X, SHAN Q, et al. Lactobacillus rhamnosus ameliorates multi-drug-resistant Bacillus cereus-induced cell damage through inhibition of NLRP3 inflammasomes and apoptosis in bovine endometritis[J].Microorganisms,2022,10（1）:137.

[14]DELEO F R. Modulation of phagocyte apoptosis by bacterial pathogens[J]. Apoptosis,2004,9（4）:399-413.

[15]WU Q, ZHU Y H, XU J, et al. Lactobacillus rhamnosus GR-1 ameliorates Escherichia coli-induced activation of NLRP3 and NLRC4 inflammasomes with differential requirement for ASC[J]. Frontiers in Microbiology,2018,9:1661.

[16]TSUCHIYA K. Switching from apoptosis to pyroptosis: Gasdermin-elicited inflammation and antitumor immunity[J]. International Journal of Molecular Sciences,2021,22（1）:426.

[17]AUBREY B J, KELLY G L, JANIC A, et al. How does p53 induce apoptosis and how does this relate to p53-mediated tumour suppression?[J]. Cell Death and Differentiation,2018,25（1）:

104-113.

[18] KIM E M, JUNG C H, KIM J, et al. The p53/p21 complex regulates cancer cell invasion and apoptosis by targeting bcl-2 family proteins[J]. Cancer Research, 2017, 77 (11): 3092-3100.

[19] ENGELAND K. Cell cycle regulation: P53-p21-RB signaling[J]. Cell Death and Differentiation, 2022, 29: 946-960.

[20] HSU S K, LI C Y, LI N I L, et al. Inflammation-related pyroptosis, a novel programmed cell death pathway, and its crosstalk with immune therapy in cancer treatment[J]. Theranostics, 2021, 11 (18): 8813-8835.

[21] PILLAI S R, KUNZE E, SORDILLO L M, et al. Application of differential inflammatory cell count as a tool to monitor udder health[J]. Journal of Dairy Science, 2001, 84 (6), 1413-1420.

[22] TRUCHET S, HONVO-HOUÉTO E. Physiology of milk secretion[J]. Best Practice and Research Clinical Endocrinology and Metabolism, 2017, 31 (4): 367-384.

[23] JIN Y, YE X, SHAO L, et al. Serum lactic dehydrogenase strongly predicts survival in metastatic na sopharyngeal carcinoma treated with palliative chemotherapy[J]. European Journal of Cancer, 2013, 49 (7): 1619-1626.

[24] WALSTON H, INES S A N, LITOVCHICK L. Dream on: Cell cycle control in development and disease[J]. Annual Review of Genetics, 2021, 55: 309-329.

[25] DE JESÚS-DÍAZ D A, MURPHY C, SOL A, et al. Host cell S phase restricts Legionella pneumophila intracellular replication by destabilizing the membrane-bound replication compartment[J]. mBio, 2017, 8 (4): e02345-e02316.

[26] PRIYA A, KAUR K, BHATTACHARYYA S, et al. Cell cycle arrest and apoptosis induced by enteroaggregative Escherichia coli in cultured human intestinal epithelial cells[J]. Journal of Medical Microbiology, 2017, 66 (2): 217-225.

[27] TAIEB F, NOUGAYRÈDE J P, WATRIN C, et al. Escherichia coli cyclomodulin Cif induces G2 arrest of the host cell cycle without

activation of the DNA-damage checkpoint-signalling pathway[J]. Cellular Microbiology,2006,8（12）: 1910-1921.

[28]YUAN L, CAI Y Q, ZHANG L, et al . Promoting apoptosis, a promising way to treat breast cancer with natural products: A comprehensive review[J]. Frontiers in Pharmacology,2022,12: 801662.

[29]ZHUANG C C, GAO J, LIU G, et al. Selenomethionine activates selenoprotein S, suppresses Fas/FasL and the mitochondrial pathway, and reduces Escherichia coli-induced apoptosis of bovine mammary epithelial cells[J]. Journal of Dairy Science, 2021,104（9）,10171-10182.

[30]JO E K, KIM J K, SHIN D M, et al.Molecular mechanisms regulating NLRP3 inflamma some activation[J]. Cellular and Molecular Immunology,2016,13（2）,148-159.

[31]SHI J J, ZHAO Y, WANG K, et al. Cleavage of GSDMD by inflammatory caspases determine s pyroptotic cell death[J]. Nature, 2015,526: 660-665.

[32]LIANG H M, WANG B, WANG J W, et al. Pyolysin of Trueperella pyogenes induces pyroptosis and IL-1β release in murine macrophages through potassium/NLRP3/caspase-1/gasdermin D pathway[J]. Frontiers in Immunology,2022,13: 832458.

[33]FRANK D, VINCE J E. Pyroptosis versus necroptosis: Similarities, differences, and crosstalk[J]. Cell Death and Differentiation,2019,26: 99-114.

[34]ZHANG L L, HOU X, SUN L C, et al. Staphylococcus aureus bacteriophage suppresses LPS-induced inflammation in MAC-T bovine mammary epithelial cells[J]. Frontiers in Immunology, 2018,9: 1614.

[35]WANG Y J, LUO W B, WANG Y F. PARP-1 and its associated nucleases in DNA damage response[J]. DNA Repair（Amst）, 2019,81: 102651.

[36]VEUGER S J, HUNTER J E, DURKACZ B W. Ionizing radiation-induced NF-κB activation requires PARP-1 function to confer

radioresistance[J]. Oncogene, 2009, 28 (6): 832-842.

[37] LIU T, ZHANG L Y, JOO D, et al. NF-κB signaling in inflammation[J]. Signal Transduction and Targeted Therapy, 2017, 2: 17023.

[38] SHEN A L, LIU L Y, HUANG Y, et al. Down-regulating HAUS6 suppresses cell proliferation by activating the p53/p21 pathway in colorectal cancer[J]. Frontiers in Cell and Developmental Biology, 2022, 9: 772077.

[39] WU G Y, LIN N, XU L N, et al. UCN-01 induces S and G2/M cell cycle arrest through the p53/p21 (waf1) or CHK2/CDC25C pathways and can suppress invasion in human hepatoma cell lines[J]. BMC Cancer, 2013, 13: 167.

[40] KIM E M, KIM J, UM H D. Bcl-2 protein targeting by the p53/p21 complex-response[J]. Cancer Research, 2018, 78 (10): 2772-2774.

[41] SAMBA-LOUAKA A, NOUGAYRÈDE J P, WATRIN C, et al. Bacterial cyclomodulin Cif blocks the host cell cycle by stabilizing the cyclin- dependent kinase inhibitors p21wafl and p27kipl[J]. Cellular Microbiology, 2008, 10 (12): 2496-2508.

[42] WANG Y J, LUO W B, WANG Y F. PARP-1 and its associated nucleases in DNA damage response[J]. DNA Repair (Amst), 2019, 81: 102651.

[43] GILBERT F B, CUNHA P, JENSEN K, et al. Differential response of bovine mammary epithelial cells to Staphylococcus aureus or Escherichia coli agonist s of the innate immune system[J]. Veterinary Researech, 2013, 44 (1): 40.

[44] CIOFU O, RIIS B, PRESSLER T, et al. Occurrence of hypermutable Pseudomonas aeruginosa in cystic fibrosis patients is associated with the oxidative stress caused by chronic lung inflammation[J]. Antimicrobial Agents and Chemotherapy, 2005, 49 (6): 2276-2282.

[45] SUN X D, JIA H D, XU Q S, et al. Lycopene alleviates H_2O_2- induced oxidative stress, inflammation and apoptosis in bovine

mammary epithelial cells via the NFE2L2 signaling pathway[J]. Food Funct, 2019,10: 6276-6285.

[46]HOU J W, ZHAO R C, XIA W Y, et al. PD-L1-mediated gasdermin C expression switches apoptosis to pyroptosis in cancer cells and facilitates tumour necrosis[J]. Nature Cell Biology, 2020,22: 1264-1275.

[47]JOHNZOH C F, DAHLBERG J, GUSTAFSON A M, et al. The effect of lipopolysaccharide-induced experimental bovine mastitis on clinical parameters, inflammatory markers, and the metabolome: A kinetic approach[J].Frontiers in Immunology, 2018,9: 1487.

[48]Zhuang C C, Zhao J H, Zhang S Y, et al. Escherichia coli infection mediate s pyroptosis via activating p53-p21 pathway-regulated apoptosis and cell cycle arrest in bovine mammary epithelial cells[J]. Microbial Pathogenesis,2023,184: 106338.

第三节 大肠杆菌感染诱导线粒体损伤的机制研究

一、概述

奶业是我国畜牧业发展的支柱性产业。奶牛乳腺炎仍然是奶牛最常见,也是最具挑战性的疾病之一。奶牛乳腺炎不仅会影响动物福利,也会造成牛奶质量下降,从而严重地影响奶业的发展并造成巨大的经济损失。奶牛乳腺炎的严重程度不仅取决于入侵病原体的种类和致病力,也受奶牛自身如哺乳期、年龄、胎次、自身免疫力以及外界饲养环境和挤奶方式等的影响,但病原微生物感染是引起奶牛乳腺炎最主要的病因。目前有150多种病原体会诱发奶牛乳腺炎。根据病原体的不同,奶牛乳腺炎可分为细菌性乳腺炎、真菌性乳腺炎、藻类性乳腺炎和支原体性乳腺炎,其中发生率最高,病情最严重且经济损失最大的是细菌性乳腺炎。根据细菌的不同来源及传播方式,奶牛乳腺炎的致病菌可以分为传染性病原菌、机会性病原菌和环境性病原菌,其中传染性病原菌主要有金黄色葡萄球菌、无乳链球菌和牛支原体等,机会性病原菌主要有表皮葡萄球菌、模拟葡萄球菌和产色葡萄球菌等,环境性病原菌主要有克雷伯杆菌、化脓隐秘杆菌和大肠杆菌等。

大肠杆菌是引起奶牛乳腺炎最常见的环境致病菌之一。Gao等人对我国161个牛场的3288份临床乳腺炎来源的牛奶样本进行分析,发现最多的病原菌是大肠杆菌,占14.4%。大肠杆菌感染会引起奶牛乳腺发生急性肿胀、疼痛和发热,乳汁呈水样、带血或凝乳性状,产奶量急剧减少,并伴有毒血症的临床症状,如体温升高、心率加快、精神沉郁和食欲不振。尽管环境乳腺炎受到管理实践的高度影响,但畜群结构和不断提高的牛奶质量标准使乳腺炎成为一种复杂的疾病,是乳制品行业的首要问题。目前,抗生素依然是国内外兽医人员治疗奶牛乳腺炎最常用的方法,然而抗生素大量且不合理的使用甚至滥用导致了耐药致病菌株的出现和传播、畜产品抗生素残留等问题,进而威胁着人类公共卫生安全。

线粒体作为细胞中重要的两层膜细胞器，不但能为细胞供给能量，还参与了调控细胞生长、分化和细胞周期、信息传递以及细胞凋亡等。大肠杆菌感染对线粒体的损伤机制仍不清晰。因此，本研究旨在进一步探究大肠杆菌感染对奶牛乳腺上皮细胞线粒体的影响及其机理，为寻找安全、高效的抗生素替代品提供新思路，为大肠杆菌性奶牛乳腺炎的防治提供新方向。

二、奶牛乳腺上皮细胞鉴定

对生长至第9代的奶牛乳腺上皮细胞进行角蛋白18免疫荧光检测，激光共聚焦显微镜拍照发现，奶牛乳腺上皮细胞胞浆中角蛋白18染成绿色，均匀分布在细胞质内，细胞核被DAPI染成蓝色，表明奶牛乳腺上皮细胞可用于后续试验。

三、奶牛乳腺上皮细胞和小鼠乳腺组织线粒体纯度鉴定

相对于细胞质部分，线粒体匀浆中高度表达了线粒体的标志物ATP5A。此外，为了确定线粒体匀浆的纯度，在对照组和大肠杆菌组线粒体匀浆中高度表达了ATP5A乳酸脱氢酶，而低表达乳酸脱氢酶（LDH，一种细胞质标记物）。这些结果表明，奶牛乳腺上皮细胞和小鼠乳腺线粒体匀浆内线粒体纯度较高，可用于后续试验。

四、大肠杆菌感染对奶牛乳腺上皮细胞超微结构的影响

利用透射显微镜观察奶牛乳腺上皮细胞线粒体的结构（图2-16），正常培养的奶牛乳腺上皮细胞内线粒体结构完整，细胞连接紧密（图2-16A），而经大肠杆菌感染的奶牛乳腺上皮细胞间隙增大，线粒体肿胀，线粒体嵴缺失且部分模糊消失（图2-16B）。这些结果表明，大肠杆菌感染造成了奶牛乳腺上皮细胞内线粒体损伤。

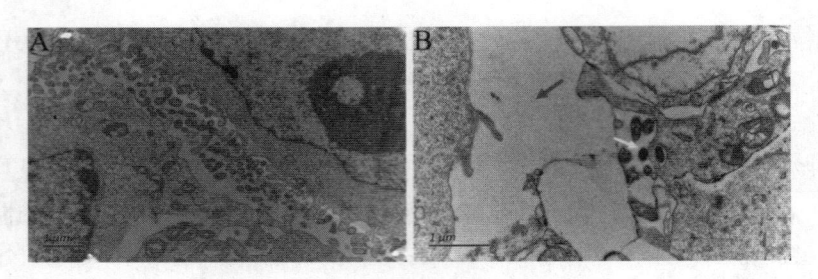

图2-16 正常奶牛乳腺上皮细胞（A）和 *E.coli* 感染奶牛乳腺上皮细胞（B）的电镜图

五、大肠杆菌感染对小鼠乳腺上病理变化的影响

组织病理结果如图2-17所示，对照组小鼠乳腺腺泡结构清晰，并有散在的乳汁；大肠杆菌组小鼠组织腺泡壁被破坏，腺泡内有游离的中性粒细胞并有散在的淡染乳汁；LPS组小鼠组织腺泡内有游离的中性粒细胞并有散在的淡染乳汁。

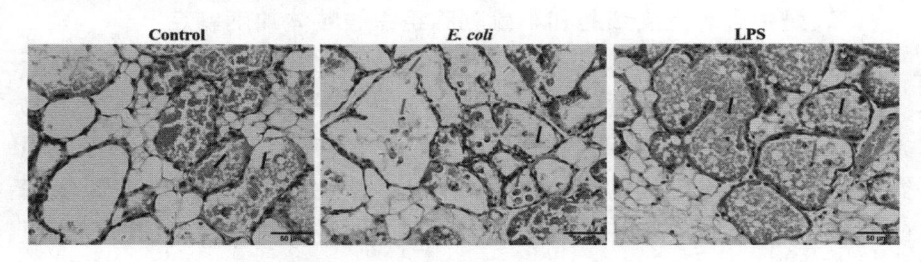

图2-17 小鼠乳腺组织病理学对比

六、大肠杆菌感染对奶牛乳腺上皮细胞和小鼠乳腺线粒体膜通透性及线粒体电子传递链的影响

相较于对照组，大肠杆菌感染或脂多糖（LPS）显著降低了奶牛乳腺上皮细胞和小鼠乳腺组织内线粒体电子传递链复合物Ⅰ、Ⅱ、Ⅲ、Ⅳ、Ⅴ活性和ATP的含量，以及奶牛乳腺上皮细胞D（520 nm）吸光值（图2-18，$P<0.01$）。大肠杆菌感染分别降低了奶牛乳腺上皮细胞D（520 nm）吸光值、线粒体电子传递链复合物Ⅰ、Ⅱ、Ⅲ、Ⅳ、Ⅴ活性和ATP含量的51.83%、40.22%、36.13%、60.67%、41.41%、57.48%和36.16%，LPS分别降低了奶牛乳腺上皮细胞D（520 nm）吸光值、线粒体电子传递链复合物Ⅰ、Ⅱ、Ⅲ、Ⅳ、Ⅴ活性和ATP含量的45.40%、38.18%、27.91%、43.42%、32.57%、43.20%和25.15%。此外，大肠杆菌感染分别降低了

小鼠乳腺线粒体电子传递链复合物Ⅰ、Ⅱ、Ⅲ、Ⅳ、Ⅴ活性和ATP含量的24.77%、71.45%、37.34%、39.38%、30.09%和27.38%，LPS分别降低了小鼠乳腺线粒体电子传递链复合物Ⅰ、Ⅱ、Ⅲ、Ⅳ、Ⅴ活性和ATP含量的19.37%、58.18%、26.74%、31.24%、32.57%、25.80%和13.11%。这些结果表明大肠杆菌感染或LPS均破坏了奶牛乳腺上皮细胞和小鼠乳腺线粒体结构的完整性，膜通透性增加，膜的流动性降低，进一步降低了线粒体传递电子的效率，最终导致线粒体能量代谢紊乱。

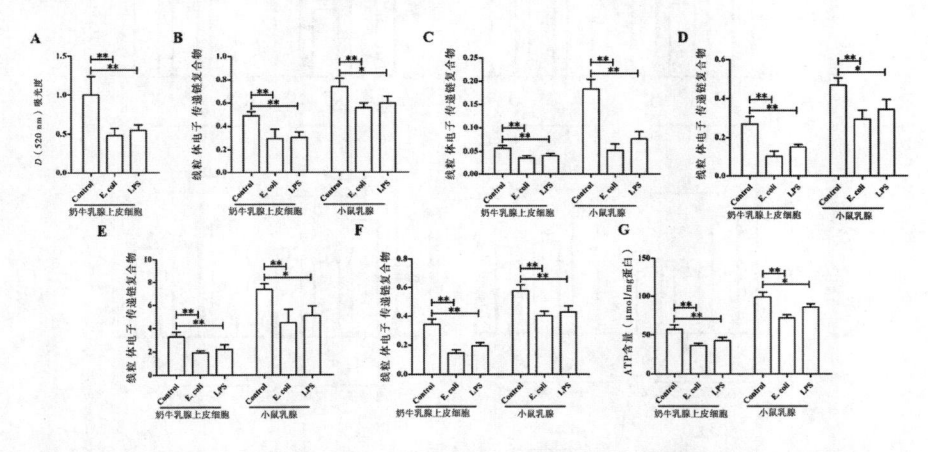

图2-18　大肠杆菌感染或LPS对奶牛乳腺上皮细胞D（520nm）吸光度（A）、电子传递链复合物（B-F）表达及ATP含量（G）的影响

七、大肠杆菌感染对奶牛乳腺上皮细胞和小鼠乳腺线粒体融合／分裂的影响

相较于对照组，大肠杆菌感染或LPS显著降低了奶牛乳腺上皮细胞和小鼠乳腺内动力蛋白相关蛋白1（$Drp1$）和线粒体分裂蛋白1（$Fis1$）调控。此外，线粒体融合蛋白1（$Mfn1$）、线粒体融合蛋白2（$Mfn2$）和蛋白质视神经萎缩1（$OPA1$）mRNA表达（图2-19，$P<0.01$）。大肠杆菌感染分别降低了奶牛乳腺上皮细胞内$Drp1$、$Fis1$、$Mfn1$、$Mfn2$和$OPA1$ mRNA表达的38.00%、42.67%、35.00%、30.00%和44.00%，LPS分别降低了奶牛乳腺上皮细胞内$Drp1$、$Fis1$、$Mfn1$、$Mfn2$和$OPA1$ mRNA表达的32.00%、33.00%、32.67%、33.00%和35.33%。此外，大肠杆菌感染分别降低了小鼠乳腺$Drp1$、$Fis1$、$Mfn1$、$Mfn2$和$OPA1$ mRNA表达的33.57%、32.20%、22.93%、31.37%和21.23%，LPS分

别降低了小鼠乳腺 $Drp1$、$Fis1$、$Mfn1$、$Mfn2$ 和 $OPA1$ mRNA 表达的 23.70%、24.20%、23.73%、30.80% 和 24.97%。这些结果表明大肠杆菌感染或 LPS 均抑制奶牛乳腺上皮细胞和小鼠乳腺内线粒体的融合与分裂。

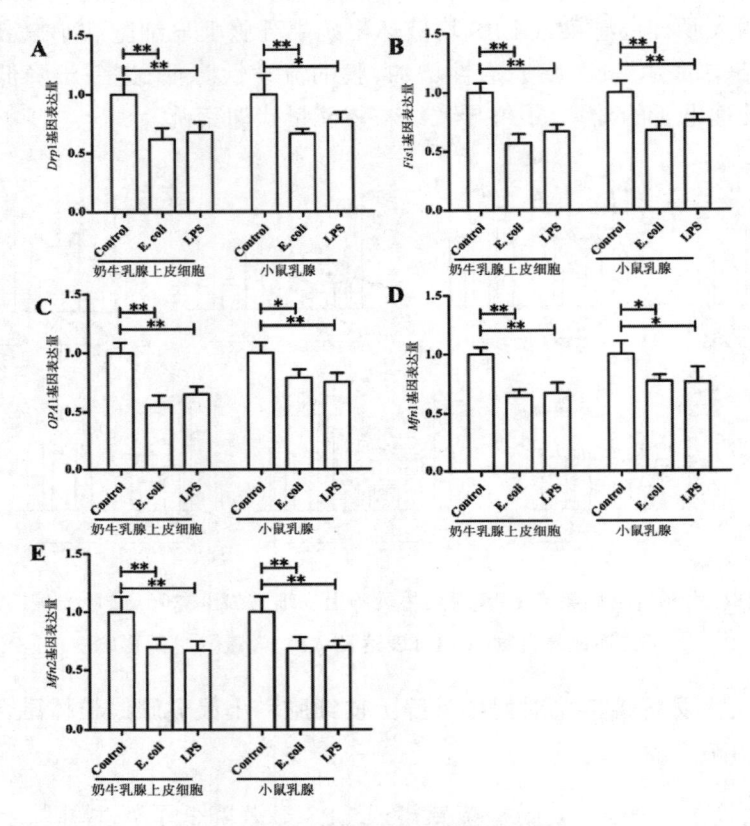

图 2-19 *E.coli* 感染或 LPS 对奶牛乳腺上皮细胞和小鼠乳腺线粒体融合/分裂的影响

八、大肠杆菌感染对奶牛乳腺上皮细胞线粒体生物发生的影响

相较于对照组,大肠杆菌感染或 LPS 显著降低了奶牛乳腺上皮细胞和小鼠乳腺内过氧化物酶增殖物激活受体 γ 共激活因子 1α（*PGC-1α*）、核呼吸因子 1（*NRF1*）和线粒体转录因子 A（*TFAM*）和 *D-Loop* 的基因表达（图 2-20, $P < 0.01$）。大肠杆菌感染分别降低了奶牛乳腺

上皮细胞内 PGC-1α、$NRF1$、$TFAM$ 和 D-$Loop$ 的基因表达的 25.93%、27.33%、31.67% 和 35.40%，LPS 分别降低了奶牛乳腺上皮细胞内 PGC-1α、$NRF1$、$TFAM$ 和 D-$Loop$ 的基因表达的 37.30%、24.33%、30.67% 和 43.07%。此外，大肠杆菌感染分别降低了小鼠乳腺 PGC-1α、$NRF1$、$TFAM$ 和 D-$Loop$ 的基因表达的 20.40%、28.50%、24.37% 和 40.13%，LPS 分别降低了小鼠乳腺 PGC-1α、$NRF1$、$TFAM$ 和 D-$Loop$ 的基因表达的 26.70%、26.60%、32.60% 和 28.27%。这些结果表明，大肠杆菌感染或 LPS 均减少了奶牛乳腺上皮细胞和小鼠乳腺内线粒体的生物发生。

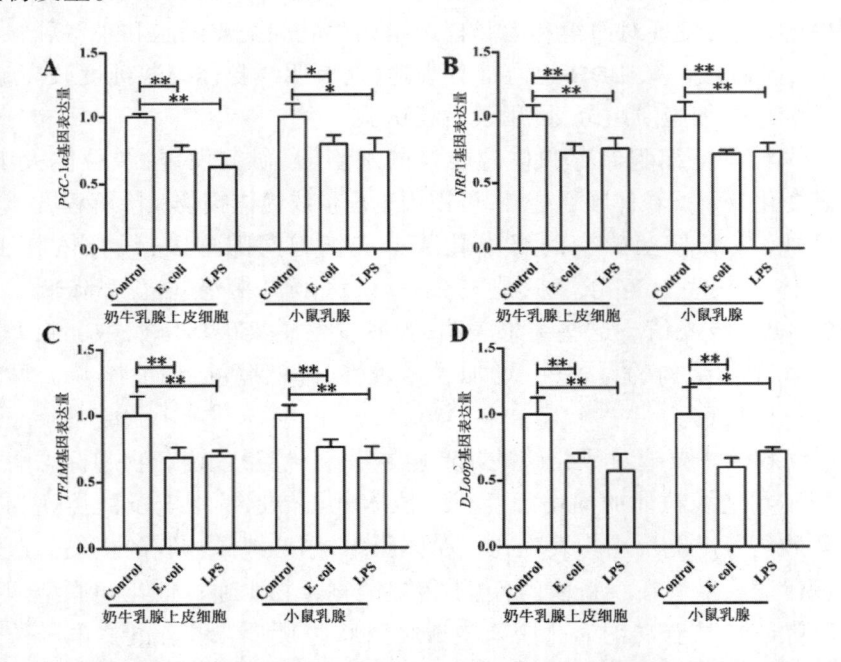

图 2-20 大肠杆菌感染或 LPS 对奶牛乳腺上皮细胞和小鼠乳腺线粒体生物发生的影响

九、讨论

奶牛乳腺上皮细胞在奶牛乳腺应对细菌感染的过程中发挥着关键作用。因此，奶牛乳腺上皮细胞已经成为研究奶牛乳腺炎发病机制的重要体外模型。角蛋白 18 是中间纤维蛋白家族中的一员，构成多种上皮细胞骨架。因此，角蛋白 18 被认为是上皮细胞特有的标志物。本研究中角蛋白18被用于鉴定奶牛乳腺上皮细胞系。研究表明，复苏后的第9代奶牛乳腺上皮细胞角蛋白18染色阳性，此研究结果与前期研究一致，表明本研究中使用的细胞系为奶牛乳腺上皮细胞。此外，在本研究中，大肠杆菌感染和LPS处理均可引起小鼠乳腺腺泡内出现游离的中性粒细胞，这表明大肠杆菌感染和LPS处理均可诱发乳腺炎。因此，本试验中小鼠乳腺炎的模型已成功建立，可用于后续试验的进行。

研究发现，ATP5A和LDH分别是线粒体和细胞质的典型标志物。本研究中，相对于细胞质部分，线粒体匀浆中ATP5A高度表达。此外，为了验证线粒体匀浆的纯度，本研究检测了线粒体匀浆中LDH蛋白的表达，发现对照组和大肠杆菌组奶牛乳腺上皮细胞和小鼠乳腺中ATP5A高度表达，LDH也有部分表达，这表明线粒体匀浆纯度较高，仅含较少量的细胞质组分，可用于后续试验。

线粒体是细胞储存和供应能量的细胞器，研究发现，线粒体大小、形态变化以及线粒体膜通透性，可被用于衡量线粒体损伤及降解程度。本研究中，大肠杆菌感染的奶牛乳腺上皮细胞间隙增大，线粒体肿胀，线粒体嵴缺失且部分模糊消失。此外，大肠杆菌感染显著降低了D(520 nm)吸光值。这些结果表明，大肠杆菌感染破坏了奶牛乳腺上皮细胞线粒体结构的完整性，增加膜通透性，降低膜的流动性，从而导致了线粒体损伤。

线粒体主要通过氧化磷酸化产生ATP，且能量代谢产物和标志酶可以用于判定线粒体的损伤程度。线粒体电子传递链中电子的传递和氧化磷酸化等产能反应均发生在线粒体膜中，因此线粒体膜的损伤会造成能量代谢紊乱。在线粒体内，电子传递链复合物Ⅰ通过催化还原型烟酰胺腺嘌呤二核苷酸的氧化脱氢生成烟酰胺腺嘌呤二核苷酸。电子传递链复合物Ⅱ通过催化琥珀酸氧化为延胡索酸，进一步使辅酶Q还原成2,6-二氯吲哚酚，最终导致O_2减少。电子传递链复合物Ⅲ是线粒体氧化磷酸化的必需物质，也是活性氧的主要来源，通过将还原型辅酶Q的氢传递给细胞色素C，生成还原型细胞色素C。电子传递链复合物Ⅳ是线粒体电子传递链的终端受体，主要通过氧化细胞色素C将O_2转化为水。电子传递链复合物Ⅴ主要是促进产生细胞所需要的ATP能量。本研究中，大肠杆菌感染显著降低了奶牛乳腺上皮细胞和小鼠乳腺内线粒体电子传递链复合物Ⅰ、Ⅱ、Ⅲ、Ⅳ、Ⅴ活性和ATP的含量，与LPS处理的结果保持一致，说明大肠杆菌感染主要通过LPS诱导线粒体膜损伤，进而干扰了线粒体的电子传递链,最终造成奶牛乳腺上皮细胞和小鼠乳腺能量代谢紊乱。

线粒体通过调控自身的分裂与融合来改变形态以满足细胞的需求。线粒体的分裂过程发生在线粒体外膜上,主要由 *Drp*1 和 *Fis*1 调控。此外,*Mfn*1、*Mfn*2 和 *OPA*1 共同介导线粒体内膜的融合。在本研究中,大肠杆菌感染或 LPS 显著降低了奶牛乳腺上皮细胞和小鼠乳腺内 *Drp*1、*Fis*1、*Mfn*1、*Mfn*2 和 *OPA*1 mRNA 表达,与 LPS 的作用一致,表明大肠杆菌感染主要通过 LPS 抑制奶牛乳腺上皮细胞和小鼠乳腺线粒体的融合与分裂。

线粒体生物发生是机体和细胞对能量需求作出的适应性反应,能够显著增加线粒体的质量。线粒体内表达的蛋白多数由线粒体 DNA 转录和翻译。线粒体 DNA 仅包含两个非编码区,非编码区无内含子,包含位移环区(D-Loop)。因此,D-Loop 可以用于衡量线粒体 DNA 的总量。本研究中,大肠杆菌感染显著降低了奶牛乳腺上皮细胞和小鼠乳腺内 *PGC*-1α、*NRF*1、*TFAM* 和 *D-Loop* 的基因表达,与 LPS 的作用保持一致,这些结果表明,大肠杆菌感染通过 LPS 减少了奶牛乳腺上皮细胞和小鼠乳腺内线粒体的生物发生。

十、结论

本研究证明了大肠杆菌感染主要通过 LPS 造成奶牛乳腺上皮细胞和小鼠乳腺线粒体能量代谢紊乱,抑制线粒体的分裂与融合,以及减少线粒体的生物发生,进而造成线粒体损伤。因此,大肠杆菌感染是通过诱导线粒体损伤造成奶牛乳腺炎,且线粒体损伤是造成乳腺损伤的重要原因之一。

参考文献

[1] SHARUN K, DHAMA K, TIWARI R, et al. Advances in therapeutic and managemental approaches of bovine mastitis: A comprehensive review[J]. The Veterinary Quarterly, 2021, 41(1): 107–136.

[2] ASHRAF A, IMRAN M. Causes, types, etiological agents, prevalence, diagnosis, treatment, prevention, effects on human health

and future aspects of bovine mastitis[J]. Animal Health Research Reviews,2020,21（1）: 36-49.

[3] SHAHEEN M, TANTARY H A. A treatise on bovine mastitis: Disease and disease economics, etiological basis, risk factors, impact on human health, therapeutic management, prevention and control strategy[J]. Advances in Dairy Research,2015,4（1）: 1.

[4] 吴琼. 乳酸杆菌减轻大肠杆菌诱发奶牛乳腺上皮细胞炎性损伤的机理 [D]. 北京: 中国农业大学,2018: 2.

[5] GAO J, BARKEMA H W, ZHANG L M, et al. Incidence of clinical mastitis and distribution of pathogens on large Chinese dairy farms[J]. Journal of Dairy Science,2017,100（6）: 4797-4806.

[6] ROUSSEL P, PORCHERIE A, RÉPÉRANT-FERTER M, et al. Escherichia coli mastitis strains: in vitro phenotypes and severity of infection in vivo[J].PLoS ONE,2017,12（7）: e0178285.

[7] MURINDA S E, IBEKWE A M, RODRIGUEZ N G, et al.Shiga toxin- Producing Escherichia coli in mastitis: An international perspective[J]. Foodborne Pathogens and Disease,2019,16（4）: 229-243.

[8] KLAAS I C, ZADOKS R N. An update on environmental mastitis: Challenging perceptions[J]. Transboundary and Emerging Diseases, 2018,65（Suppl 1）: 166-185.

[9] DONG L, MENG L, LIU H M, et al. Effect of therapeutic administration of β-lactam antibiotics on the bacterial community and antibiotic resistance patterns in milk[J]. Journal of Dairy Science,2021,104（6）: 7018-7025.

[10] PILLAI S R, KUNZE E, SORDILLO L M, et al. Application of differential inflammatory cell count as a tool to monitor udder health[J]. Journal of Dairy Science,2001,84（6）: 1413-1420.

[11] TRUCHET S, HONVO-HOUÉTO E. Physiology of milk secretion[J]. Best Practice and Research Clinical Endocrinology and Metabolism,2017,31（4）: 367-384.

[12] XU F B, LIU Y F, ZHAO H S, et al. Aluminum chloride caused liver dysfunction and mitochondrial energy metabolism disorder

in rat[J]. Journal of Inorganic Biochemistry, 2017, 174: 55-62.

[13] 马国源. 低剂量亚硝酸钠抑制牦牛肉肌红蛋白氧化的作用机制 [D]. 兰州: 甘肃农业大学, 2021: 73-75.

[14] GRIFFITHS E J.Mitochondria-potential role in cell life and death[J].Cardiovascular Research, 2000, 46 (1): 24-27.

[15] 吕宝北, 赵鹏翔, 张鑫, 等. 线粒体呼吸链复合物Ⅰ结构和功能的研究进展 [J]. 现代生物医学进展, 2018, 18 (2): 356-359, 380.

[16] 马艳艳, 杨艳玲. 线粒体呼吸链复合物Ⅱ缺陷与疾病 [J]. 中国当代儿科杂志, 2012, 14 (10): 723-727.

[17] 崔树娜, 钱静, 卜平. 线粒体复合体Ⅲ抑制剂抗霉素 A 对巨噬细胞免疫功能的影响 [J]. 中国药理学与毒理学杂志, 2015, 29 (4): 573-577.

[18] ZHAN G K, GUO M Y, LI Q G, et al. Drp1-dependent mitochondrial fission mediates corneal injury induced by alkali burn[J]. Free Radical Biology and Medicine, 2021, 176: 149-161.

[19] MILANOWSKI P, KOSIOR-JARECKA E, ŁUKASIK U, et al. Associations between OPA1, MFN1, and MFN2 polymorphisms and primary open angle glaucoma in Polish participants of European ancestry[J]. Ophthalmic Genetics, 2022, 43 (1): 42-47.

[20] KATHIRAVAN P, KATARIA R S, MISHRA B P, et al. Population structure and phylogeography of Toda buffalo in Nilgiris throw light on possible origin of aboriginal Toda tribe of South India[J]. Journal of Animal Breeding and Genetics, 2011, 128 (4): 295-304.

[21] LIN J D, PUIGSERVER P, DONOVAN J, et al. Peroxisome proliferator-activated receptor γ coactivator 1β (PGC-1β), A novel PGC-1-related transcription coactivator associated with host cell factor[J]. Journal of Biological Chemistry, 2002, 277 (3): 1645-1648.

第四节 大肠杆菌感染诱导乳腺上皮细胞铁死亡的机制研究

一、概述

乳腺炎是奶牛最常见和最具挑战性的疾病之一,对动物福利和营利能力以及牛奶质量都有有害影响。大肠杆菌是导致奶牛乳腺炎最常见的环境病原体。在牛乳腺中,大肠杆菌感染会引起急性肿胀、发热和疼痛,牛奶呈水样、血性或凝乳,产奶量减少,并伴有毒血症的临床症状,包括体温和心率升高和食欲不振。尽管环境乳腺炎受环境和管理的高度影响,但畜群规模的增加和加工商对牛奶质量的标准使乳腺炎成为一种复杂的疾病,这仍然是乳制品行业的首要问题。因此,对大肠杆菌乳腺炎的进一步研究是当务之急。

大肠杆菌在牛乳腺中增殖,导致乳腺炎的发生。自噬是一种程序性细胞死亡方式,作为一种降解途径,在消除过多的错误折叠蛋白质和受损的细胞器聚集体方面发挥着关键作用。ROS 的积累通过增加脂质过氧化和降低抗氧化酶的活性来促进牛乳腺上皮细胞自噬依赖性细胞死亡。奶牛乳腺炎的发病机制可能与自噬的稳态功能障碍有关,因此探讨奶牛乳腺炎细胞自噬的调控机制至关重要。线粒体自噬是选择性地消除老化和受损线粒体的过程。PINK1-Parkin 介导的通路是线粒体自噬的主要调节因子。然而,大肠杆菌感染对奶牛乳腺上皮细胞中线粒体自噬的影响,相关文献记载并不充分。

铁死亡是一种由铁依赖性脂质过氧化引起的程序性细胞死亡方式。然而,铁代谢失调涉及多种生理和病理过程,包括癌细胞死亡、神经毒性、神经退行性疾病、急性肾功能衰竭、药物性肝毒性、肝脏和心脏缺血/再灌注损伤、细胞免疫和乳腺上皮细胞功能障碍。铁离子(Fe^{3+})由转铁蛋白受体 1(TfR1)输入并沉积在核内体中,在核内体中,铁离子通过前列腺 3 的六跨膜上皮抗原(STEAP3)进行转化。随后,Fe^{2+} 通过二价金属转运体1（DMT1）从核内体释放到细胞质中,通过Fenton反应提高ROS水平和铁沉积。此外,由二肽胱氨酸产生的半胱氨酸通过细胞表面胱氨酸/谷氨酸反转运系统Xc-（由SLC3A2和SLC7A11二聚体组成）转运到细胞内,促进谷胱甘肽（GSH）和谷胱甘肽过氧化物酶4GPX4的合成,从而抑制铁致细胞死亡。此外,过量的铁被储存在铁蛋白中,铁蛋白是一种铁储存蛋白复合物,包括铁蛋白轻链和铁蛋白重链1。铁蛋白的降解需要货物受体核受体共激活剂4（NCOA4）将生物可利用的铁释放回细胞质。然而,自噬和铁死亡之间的相互作用及导致这两种细胞死亡的潜在机制尚未完全阐明。

Wnt/β-catenin信号通路在调节铁死亡中具有重要作用。Wnt配体结合卷曲受体,通过由大肠腺瘤性息肉病蛋白、轴素、酪蛋白激酶1和糖原合酶激酶3β（GSK3β）组成的破坏复合物促进β-catenin转运进入细胞核,进而增强c-MyC并抑制P62表达,从而促进自噬和铁凋亡。然而,关于大肠杆菌感染对奶牛乳腺上皮细胞中Wnt/β-catenin通路调节的自噬和铁死亡的影响的数据有限。本研究采用大肠杆菌感染奶牛乳腺上皮细胞体外实验模型,研究大肠杆菌感染对奶牛乳腺上皮细胞Wnt/β-catenin通路调控的自噬和铁凋亡的影响,为预防和管理大肠杆菌乳腺炎提供新的见解,并确定治疗乳腺炎的潜在非抗生素替代疗法。

二、细胞角蛋白18和大肠杆菌感染对细胞铁死亡的影响

奶牛乳腺上皮细胞传6代后,通过评价上皮细胞特异性细胞角蛋白18的表达来确定奶牛乳腺上皮细胞的纯度。复苏的奶牛乳腺上皮细胞显示出强烈的角蛋白18阳性染色（图2-21A）。与对照组相比,大肠杆菌感染或埃拉斯汀（erastin）显著增加了LDH活性（$P < 0.01$,图2-21B）。在大肠杆菌感染和erastin处理的奶牛乳腺上皮细胞中,线粒体变小,线粒体膜密度增加,线粒体嵴减少或消失（图2-21C）。与对照组相比,大肠杆菌感染或erastin使Fe^{2+}浓度显著升高（$P < 0.01$,图2-21D）。

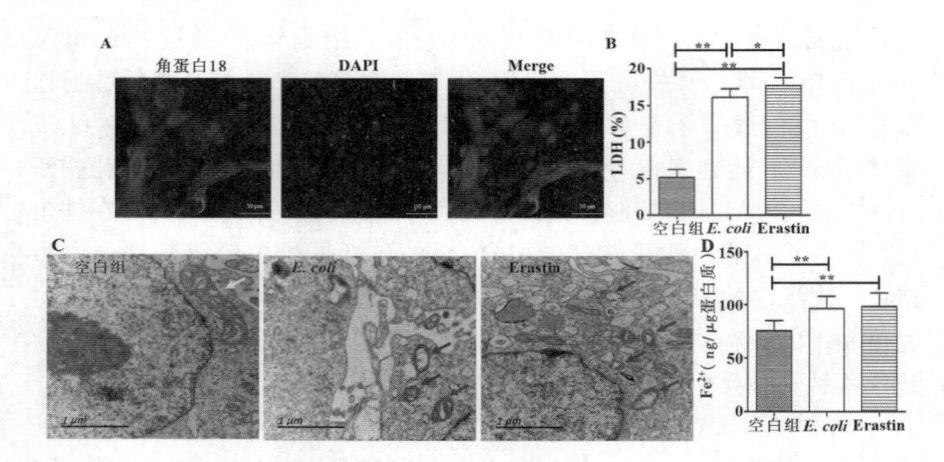

图 2-21 奶牛乳腺上皮细胞纯度、LDH 活性和大肠杆菌感染诱导的奶牛乳腺上皮细胞铁铁死亡

三、大肠杆菌感染引起奶牛乳腺上皮细胞发生氧化应激

与对照组相比，大肠杆菌组和 erastin 组的细胞内 ROS 生成（图 2-22A 和 B）以及 H_2O_2（图 2-22C）、4-羟基壬烯醛（4-HNE）（图 2-22D）和 MDA（图 2-22E）浓度显著高于对照组（$P < 0.01$）。此外，与对照组相比，大肠杆菌或 erastin 处理的奶牛乳腺上皮细胞中总抗氧化能力（T-AOC）（图 2-22F）和过氧化氢酶 CAT（图 2-22H）活性和 GSH 浓度（图 2-22G）均显著下降（$P < 0.05$；$P < 0.01$）。

四、大肠杆菌感染抑制系统 Xc-/GPX4 轴

与对照组相比，大肠杆菌感染或 erastin 显著抑制了（$P < 0.01$）SLC7A11（图 2-23A 和 B）、GPX4（图 2-23A 和 C）和铁蛋白重链（FTH1）（图 2-23A 和 D）蛋白表达，而大肠杆菌感染或 erastin 显著增加了（$P < 0.01$）NCOA4（图 2-23A 和 E）蛋白表达。

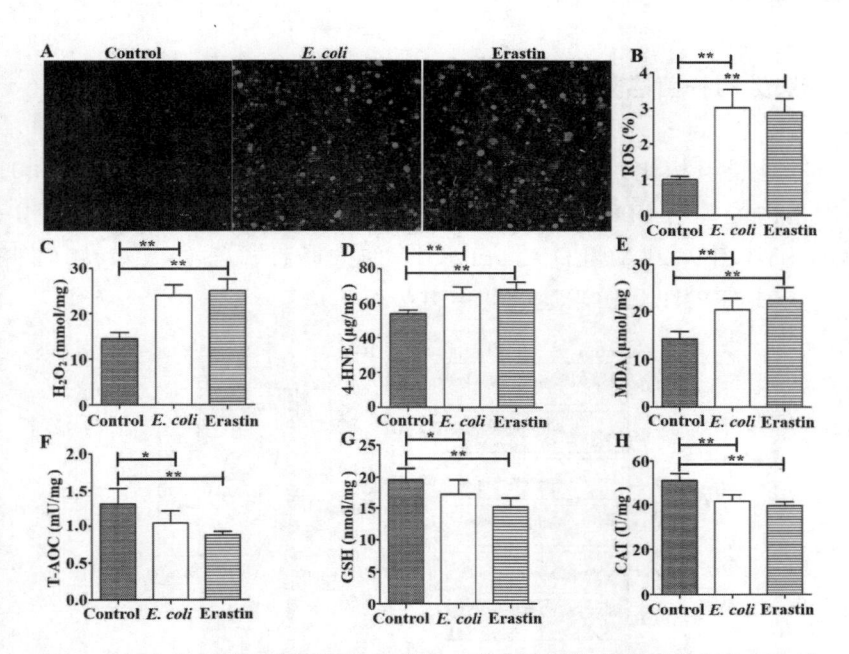

图 2-22 大肠杆菌感染诱导奶牛乳腺上皮细胞氧化应激并抑制抗氧化酶

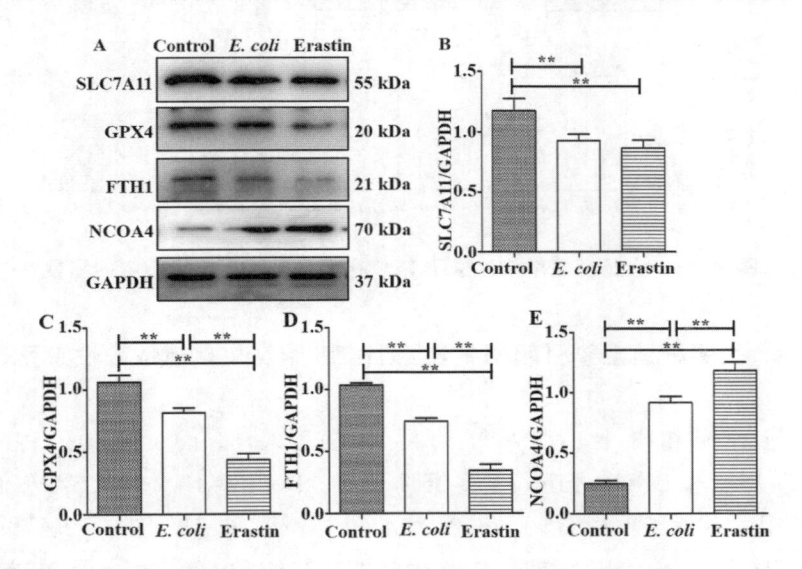

图 2-23 大肠杆菌感染抑制奶牛乳腺上皮细胞中的系统 Xc-/GPX4 轴

五、大肠杆菌感染引起奶牛乳腺上皮细胞铁代谢失衡

与对照组相比,大肠杆菌感染或 erastin 处理显著增加($P < 0.01$)转铁蛋白受体(p-TFR)(图 2-24A 和 B)、DMT1(图 2-24A 和 C)和 STEAP3(图 2-24A 和 D)蛋白水平,而大肠杆菌感染或 erastin 抑制($P < 0.01$)FPN1 蛋白表达(图 2-24A 和 E)。

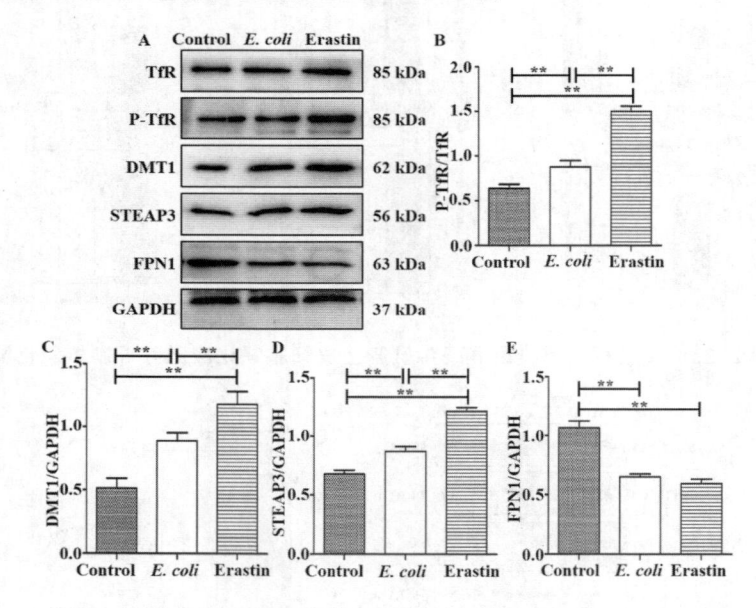

图 2-24　大肠杆菌感染促进了 Fe^{2+} 在奶牛乳腺上皮细胞中的积累

六、大肠杆菌感染引起奶牛乳腺上皮细胞发生自噬依赖性铁死亡

与对照组相比,大肠杆菌感染显著增加了($P < 0.01$)PINK1(图 2-25A 和 B)、Parkin(图 2-25A 和 C)、Beclin-1(图 2-25A 和 D)、自噬相关基因5(ATG5)(图 2-25A 和 E)和 LC3II(图 2-25A 和 G)蛋白表达,以及细胞内 ROS 产生(图 2-25H 和 I)、Fe^{2+}(图 2-25J)、H_2O_2(图 2-25K)、4-HNE(图 2-25L)和 MDA(图 2-25M)含量,而显著抑制了($P < 0.01$)P62(图 2-25A 和 F)蛋白表达、T-AOC(图 2-25N)、CAT(图 2-25P)活性以及 GSH 含量(图 2-25O)。此外,氯喹作为一种阻止溶酶体自噬降解的溶酶体抑制剂显著减弱($P < 0.05$;

$P <0.01$)这种变化(图 2-25A-P)。相反,雷帕霉素促进($P < 0.05$;$P <0.01$)大肠杆菌感染诱导线粒体自噬依赖性铁死亡激活增强($P < 0.05$;$P <0.01$,图 2-25A-P)。

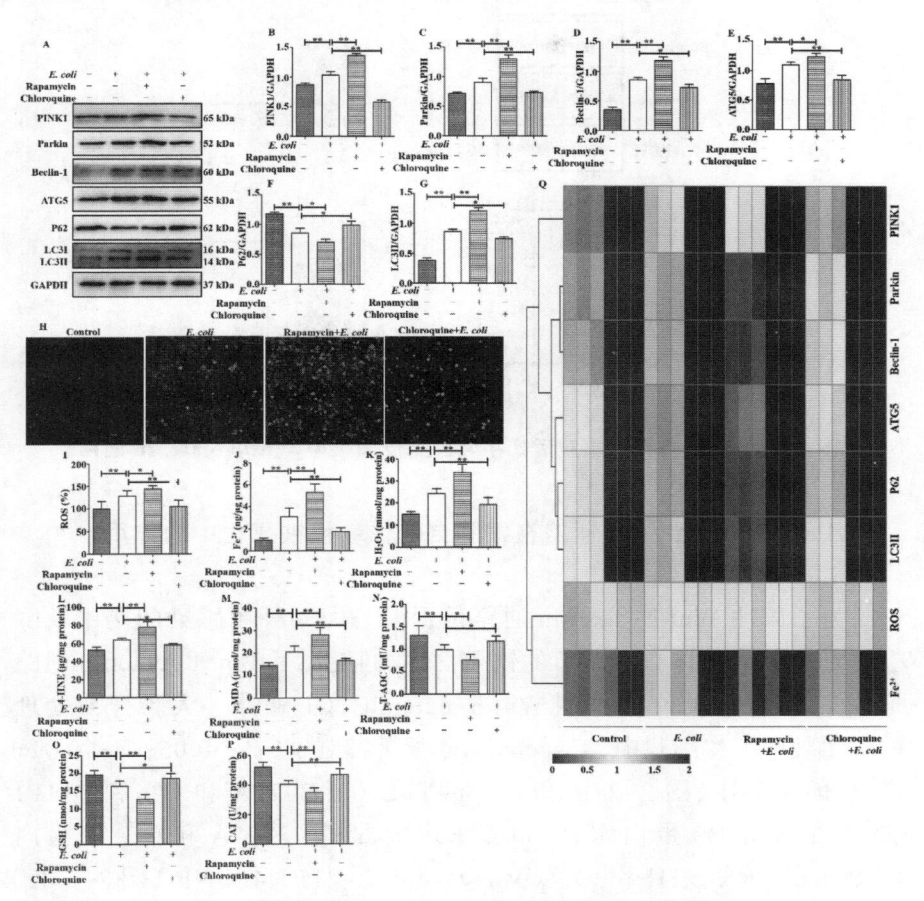

图 2-25 大肠杆菌感染引起奶牛乳腺上皮细胞自噬依赖性铁死亡

七、大肠杆菌感染激活 Wnt/β-catenin 通路

大肠杆菌感染组 p-GSK3β(图 2-26A 和 B)、β-catenin(图 2-26A 和 C)和 c-MyC(图 2-26A 和 D)蛋白表达水平均高于对照组($P < 0.01$)。

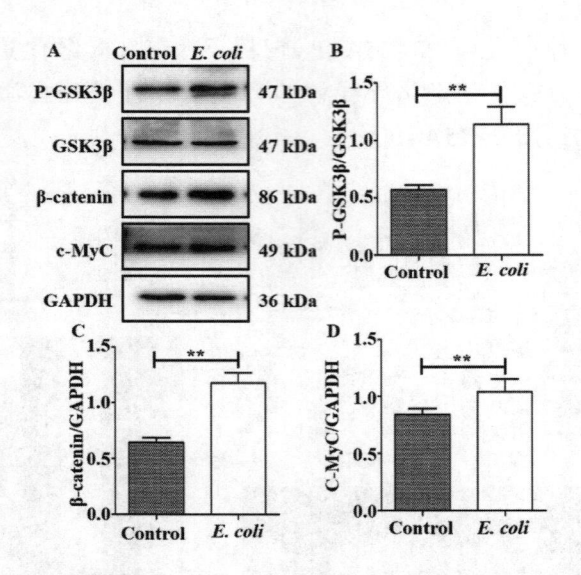

图 2-26　大肠杆菌感染激活奶牛乳腺上皮细胞中 Wnt/β-catenin 通路

八、Wnt/β-catenin 通路在细胞自噬依赖性铁死亡中的作用

为了阐明 Wnt/β-catenin 通路是否参与大肠杆菌感染的奶牛乳腺上皮细胞中线粒体自噬调节的铁死亡，我们使用了 Wnt/β-catenin 通路抑制剂 foscenvivint 来抑制 Wnt/β-catenin/TCF 通路。大肠杆菌处理的奶牛乳腺上皮细胞中，foscenvivint 显著抑制了（$P < 0.05$；$P < 0.01$）增加的 LC3II（图 2-27A 和 C）、p-TFR（图 2-27A 和 D）和 DMT1（图 2-27A 和 E）蛋白水平，以及 Fe^{2+} 含量（图 2-27A 和 G）、细胞内 ROS 生产（图 2-27H 和 I），以及 H_2O_2（图 2-27J）、4-HNE（图 2-27K）和 MDA（图 2-27L）含量，并缓解了 P62 的下降（图 2-27A 和 B）、GPX4（图 2-27A 和 F）蛋白表达、T-AOC（图 2-27）和 CAT（图 2-27O）活性和 GSH 含量（图 2-27N）。

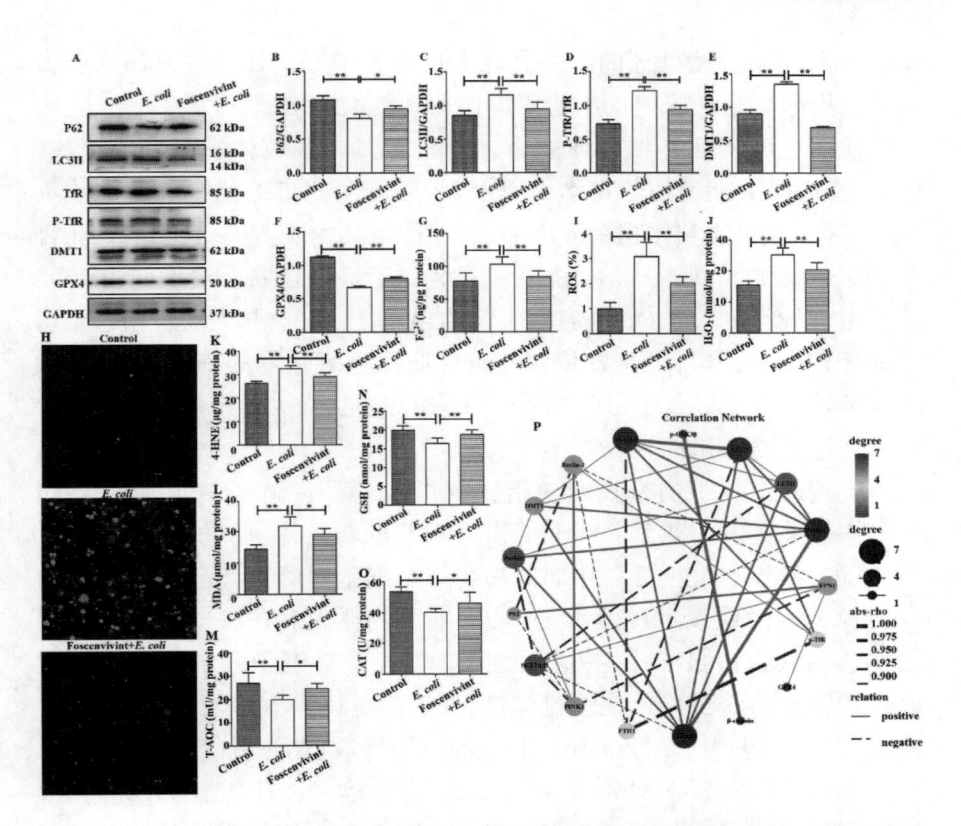

图 2-27　抑制 Wnt/β-catenin 通路可逆转大肠杆菌感染诱导的奶牛乳腺上皮细胞自噬依赖性铁死亡

九、讨论

众所周知,奶牛乳腺上皮细胞对细菌感染有反应,这使其成为研究乳腺炎发病机制的理想体外模型。因此,我们利用奶牛乳腺上皮细胞来探讨大肠杆菌感染是否通过 Wnt/β-catenin 途径介导铁死亡。在本研究中,通过 CK18(管腔上皮细胞的标记物)染色证实了奶牛乳腺上皮细胞的纯度,奶牛乳腺上皮细胞呈强烈阳性染色,与我们之前的研究结果一致。LDH 作为糖酵解途径的关键酶,催化丙酮酸生成乳酸的可逆反应,同时伴有 NADH 和 NAD^+ 的相互转化。LDH 含量丰富,存在于大多数细胞类型中。此外,LDH 释放是评估细胞膜完整性的敏感生物指标,LDH 释放增加表明细胞膜通透性增强,细胞膜受损。因此,在

erastin 和大肠杆菌处理的奶牛乳腺上皮细胞中，LDH 活性升高表明大肠杆菌引起奶牛乳腺上皮细胞膜损伤并抑制细胞增殖。

铁死亡的特征是线粒体变小，膜密度增加，线粒体嵴减少或消失，ROS 和 Fe^{2+} 浓度增加。在本研究中，大肠杆菌感染的奶牛乳腺上皮细胞线粒体变小，线粒体膜密度增加，线粒体嵴减少或消失。此外，大肠杆菌感染（MOI=5）或 erastin 处理显著增加了奶牛乳腺上皮细胞内 ROS 的产生，以及 Fe^{2+}、H_2O_2、4-HNE 和 MDA 的浓度，这与大肠杆菌感染诱导铁死亡一致，其特征是铁依赖性致死脂质过氧化物的积累。同样，大肠杆菌悬浮液可引起小鼠间充质干细胞的铁死亡，大肠杆菌孵育可引起草鱼红细胞的铁死亡。

缺氧癌细胞中的铁代谢包括铁的摄入、利用、储存和外排，这些过程受铁相关蛋白的调节。DMT1 作为 Fe^{2+} 的转运体，促进 Fe^{2+} 的摄取，STEAP3 有助于铁的摄取并维持铁在多个细胞中的储存。此外，FTH1 是一种关键的铁死亡调节因子，其水平影响体内和体外对铁死亡的易感性。在铁蛋白自噬中，货物受体 NCOA4 直接识别并结合 FTH1，然后将铁结合的铁蛋白传递给自噬体，使溶酶体降解并释放铁，从而诱导铁凋亡。本研究中，在 erastin 和大肠杆菌处理的奶牛乳腺上皮细胞中，P-TFR、DMT1、STEAP3 和 NCOA4 蛋白表达显著升高，FTH1 蛋白表达水平显著降低，提示大肠杆菌感染增加了 Fe^{2+} 的积累。此外，FPN1 是一种膜蛋白，通过从细胞质输出铁离子参与细胞内铁稳态和抗铁凋亡作用。因此，在 erastin 和大肠杆菌处理的奶牛乳腺上皮细胞中，FPN1 蛋白水平下降表明大肠杆菌感染破坏了铁代谢，并通过抑制 Fe^{2+} 的转移来促进 Fe^{2+} 的积累。

Xc-/GPX4 轴在预防脂质过氧化介导的铁凋亡中起重要作用。在本研究中，大肠杆菌感染或 erastin 处理显著抑制奶牛乳腺上皮细胞中 SLC7A11 和 GPX4 的蛋白表达以及 GSH 浓度。同样，脓毒症相关脑病通过抑制小鼠 Xc-/GPX4 轴引发海马铁死亡。因此，我们推断大肠杆菌感染通过抑制 Xc-/GPX4 轴诱导奶牛乳腺上皮细胞铁死亡。

持续的铁过载可能导致线粒体损伤，随后激活线粒体自噬，从而为脂质过氧化或激活线粒体依赖的恶性循环提供额外的铁来源，并直接释放铁、ROS 或过氧化脂质，从而触发和放大铁死亡。线粒体自噬通过去除老化的、受损的，以及因此可能具有细胞毒性的线粒体，对维持线粒体稳态至关重要。据报道，PINK1/Parkin 通路启动和调节线粒体自噬,从而识别和消除受损的线粒体。生理条件下,PINK1 被导入线粒体,锚定在线粒体内膜上,并被线粒体蛋白酶降解。然而,在受损的线粒体中,PINK1 聚集在线粒体外膜上,招募并激活 Parkin,通过降解 p62/SQSTM1 自噬受体与自噬体上的 LC3 结合,并促进受损线粒体的消除。此外,ATG5 对于自噬小泡的形成是必不可少的,Beclin-1 可以介导其他自噬蛋白向吞噬细胞的定位,从而调节哺乳动物自噬小体的形成和成熟。在本研究中,大肠杆菌处理的奶牛乳腺上皮细胞中,PINK1、Parkin、Becline-1、ATG5 和 LC3II 蛋白表达显著升高,P62 蛋白表达水平显著降低,雷帕霉素(一种自噬激活剂)或氯喹(一种阻止溶酶体自噬降解的溶酶体抑制剂)显著促进了 P62 蛋白的表达,提示临床乳腺炎奶牛分离的大肠杆菌(MOI = 5)感染可引起 PINK/parkin 介导的线粒体自噬。相比之下,大肠杆菌菌株(血清型 O111:K58,CVCC1450)感染(2×10^7CFU 或 MOI = 66)6 h 或 8 h 后,猪乳腺上皮细胞或奶牛乳腺上皮细胞的线粒体自噬被抑制。这种差异可能是由于致病性或感染持续时间的差异所致。此外,雷帕霉素(一种自噬激活剂)显著增加大肠杆菌处理的奶牛乳腺上皮细胞细胞内 ROS 生成、Fe^{2+}、H_2O_2、4-HNE 和 MDA 浓度,抑制 GSH 浓度、T-AOC 和 CAT 活性,而氯喹(一种自噬抑制剂)处理的效果相反。因此,我们推断大肠杆菌感染通过激活 PINK/Parkin 介导的线粒体自噬诱导奶牛乳腺上皮细胞铁死亡。

有趣的是,Wnt/β-catenin/TCF 途径抑制剂 foscenvivint 显著抑制了大肠杆菌处理的奶牛乳腺上皮细胞中 Fe^{2+}、H_2O_2、4-HNE 和 MDA 浓度,细胞内 ROS 生成,以及 LC3II、P-TFR 和 DMT1 的表达,显著促进了 P62 和 GPX4 表达,T-AOC 和 CAT 活性,以及 GSH 浓度。Wnt/β-catenin 通路的激活通过增加 c-MyC 转录促进铁摄取来诱导 TFR1 和 DMT1 表达。此外,β-catenin/TCF4 转录复合体直接结合 GPX4 的启动子区域并诱导其表达,从而抑制铁致细胞死亡。此外,β-catenin 通过 TCF4 抑制 P62 的转录。此外,在本研究的相关网络中,PINK1、Parkin、Beclin-1、ATG5、P62、LC3II、SCL7A11、GPX4、FTH1、P-TFR、NCOA4、FPN1、DMT1、STEAP3、p-GSK3β、β-catenin、c-MyC 蛋白之间也相互存在正调控或负调控关系。因此,我们得出结论:大肠杆菌感染通过激活 Wnt/β-catenin 途径诱导铁死亡—线粒体自噬增加,降低 GPX4 表达。

十、结论

大肠杆菌感染通过激活 Wnt/β-catenin 途径诱导铁死亡,促进线粒体自噬,抑制奶牛乳腺上皮细胞中 GPX4 的表达(图 2-28)。此外,我们推断,在大肠杆菌诱导的乳腺炎中,通过抑制 Wnt/β-catenin 途径和线粒体自噬来下调铁死亡的治疗策略可能有助于保留牛乳腺功能。

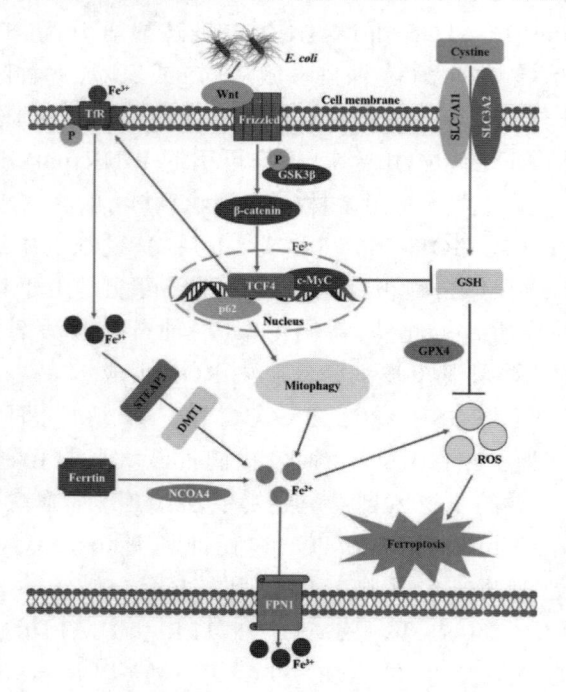

图 2-28 大肠杆菌感染诱导的铁死亡的信号通路

参考文献

[1] SHARUN K, DHAMA K, TIWARI R, et al. Advances in therapeutic and managemental approaches of bovine mastitis: A comprehensive review[J]. The Veterinary Quarterly, 2021, 41(1): 107-136.

[2] GAO J, BARKEMA H W, ZHANG L M, et al. Incidence of clinical mastitis and distribution of pathogens on large Chinese dairy farms[J]. Journal of Dairy Science, 2017, 100(6): 4797-4806.

[3] ROUSSEL P, PORCHERIE A, RÉPÉRANT-FERTER M, et al. Escherichia coli mastitis strains: in vitro phenotypes and severity of infection in vivo[J]. PLoS One, 2017, 12（7）: e0178285.

[4] MURINDA S E, IBEKWE A M, RODRIGUEZ N G, et al. Shiga toxin-Producing Escherichia coli in mastitis: An international perspective[J]. Foodborne Pathogens and Disease, 2019, 16（4）: 229-243.

[5] KLAAS I C, ZADOKS R N. An update on environmental mastitis: Challenging perceptions[J]. Transboundary and Emerging Disease, 2017, 65: 166-185.

[6] PETZL W, ZERBE H, GÜNTHER J, et al. Pathogen-specific responses in the bovine udder. Models and immunoprophylactic concepts[J]. Research in Veterinary Science, 2018, 116: 55-61.

[7] FILOMEN I G, DE ZIO D, CECCONI F. Oxidative stress and autophagy: The clash between damage and metabolic needs[J]. Cell Death and Differentiation, 2015, 22（3）: 377-388.

[8] CHEN P, YANG J Y, WU N W, et al. Streptococcus lutetiensis induces autophagy via oxidative stress in bovine mammary epithelial cells[J]. Oxidative Medicine and Cellular Longevity, 2022, 2022: 2549772.

[9] SU L J, ZHANG J H, GOMEZ H, et al. Mitochondria ROS and mitophagy in acute kidney injury[J]. Autophagy, 2023, 19（2）: 401-414.

[10] YAMADA T, DAWSON T M, YANAGAWA T, et al. SQSTM1/p62 promotes mitochondrial ubiquitination independently of PINK1 and PRKN/parkin in mitophagy[J]. Autophagy, 2019, 15（11）: 2012-2018.

[11] XIE Y, HOU W, SONG X, et al. Ferropto s is: Process and function[J]. Cell Death and Differentiation, 2016, 23（3）: 369-379.

[12] ZHU G Q, SUI S P, SHI F Y, et al. Inhibition of USP14 suppresses ferropto s is and inflammation in LPS-induced goat mammary epithelial cells through ubiquitylating the IL-6

protein[J]. Hereditas, 2022, 159（1）: 21.

[13] MASALDAN S, BUSH A I, DEVOS D, et al. Striking while the iron is hot: Iron metabolism and ferroptosis in neurodegeneration[J]. Free Radical Biology and Medicine, 2019, 133: 221–233.

[14] LI D S, LI Y S. The interaction between ferropto sis and lipid metabolism in cancer[J]. Signal Transduction and Targeted Therapy, 2020, 5（1）: 108.

[15] YU F, ZHANG Q P, LIU H Y, et al. Dynamic O-GlcNAcylation coordinates ferritinophagy and mitophagy to activate ferroptosis[J]. Cell Discovery, 2022, 8（1）: 40.

[16] DOWDLE W E, NYFELER B, NAGEL J, et al. Selective VPS34 inhibitor blocks autophagy and uncovers a role for NCOA4 in ferritin degradation and iron homeostasis in vivo[J]. Nature Cell Biology, 2014, 16: 1069–1079.

[17] TORTI S V, TORTI F M. Iron and cancer: More ore to be mined[J]. Nature Reviews Cancer, 2013, 13（5）: 342–355.

[18] NUSSE R, CLEVERS H. Wnt/β–catenin signaling, disease, and emerging therapeutic modalities[J]. Cell, 2017, 169（6）: 985–999.

[19] XIONG X P, HASANI S, YOUNG L E A, et al. Activation of Drp1 promotes fatty acids–induced metabolic reprograming to potentiate Wnt signaling in colon cancer[J]. Cell Death and Differentiation, 2022, 29（10）: 1913–1927.

[20] PILLAI S R, KUNZE E, SORDILLOL M, et al. Application of differential inflammatory cell count as a tool to monitor udder health[J]. Journal of Dairy Science, 2001, 84（6）: 1413–1420.

[21] TRUCHET S, HONVO-HOUÉTO E. Physiology of milk secretion[J]. Best Practic and Research Clinical Endocrinology and Metabolism, 2017, 31（4）: 367–384.

[22] LIU Y, DENG Z J, XU S Y, et al. Mycoplasma bovis subverts autophagy to promote intracellular replication in bovine mammary epithelial cells cultured in vitro[J]. Veterinary Research, 2021, 52

(1): 130.

[23] CHENG A X, ZHANG P, WANG B, et al. Aurora–A mediated phosphorylation of LDHB promotes glycolysis and tumor progression by relieving the substrate–inhibition effect[J]. Nature Communications, 2019, 10: 5566.

[24] JIN Y, YE X, SHAO L, et al. Serum lactic dehydrogenase strongly predicts survival in metastatic nasopharyngeal carcinoma treated with palliative chemotherapy[J]. European Journal of Cancer, 2013, 49 (7): 1619–1626.

[25] SHAHID M, WANG J F, GU X L, et al. Prototheca zopfii induced ultrastructural features associated with apoptosis in bovine mammary epithelial cells[J]. Frontiers in Cellular and Infection Microbiology, 2017, 7: 299.

[26] ZHANG C, LIU X Y, JIN S D, et al. Ferroptosis in cancer therapy: A novel approach to reversing drug resistance[J]. Molecular Cancer, 2022, 21 (1): 47.

[27] JIANG L, KON N, LI T Y, et al. Ferroptosis as a p53–mediated activity during tumour suppression[J]. Nature, 2015, 520 (7545): 57–62.

[28] PAN Y C, LI J M, WANG J L, et al. Ferroptotic MSCs protect mice against sepsis via promoting macrophage efferocytosis[J]. Cell Death and Disease, 2022, 13: 825.

[29] YANG M X, LU Z J, LI F L, et al. Escherichia coli induced ferroptosis in red blood cells of grass carp (Ctenopharyngodon idella)[J]. Fish and Shellfish Immunology, 2021, 112: 159–167.

[30] TORTI S V, TORTI F M. Iron: The cancer connection[J]. Molecular Aspects of Medicine, 2020, 75: 100860.

[31] MANATSCHAL C, PUJOL-GIMÉNEZ J, POIRIER M, et al. Mechanistic basis of the inhibition of SLC11/NRAMP–mediated metal ion transport by bis–isothiourea substituted compounds[J]. elife, 2019, 8: e51913.

[32] YAN Y L, LIANG Q J, XU Z J, et al. Downregulated ferropto sis-

related gene steap3 as a novel diagnostic and prognostic target for hepatocellular carcinoma and its roles in immune regulation[J]. Frontiers in Cell and Developmental Biology, 2021, 9: 743046.

[33] RUI T Y, WANG H C, LI Q Q, et al. Deletion of ferritin H in neurons counteracts the protective effect of melatonin against traumatic brain injury-induced ferroptosis[J]. Journal of Pineal Research, 2021, 70 (2): e12704.

[34] FANG Y Y, CHEN X C, TAN Q Y, et al. Inhibiting ferroptosis through disrupting the NCOA4-FTH1 interaction: A new mechanism of action[J]. ACS Central Science, 2021, 7 (6): 980-989.

[35] BILLESBØLLE C B, AZUMAYA C M, KRETSCH R C, et al. Structure of hepcidin-bound ferroportin reveals iron homeostatic mechanisms[J]. Nature, 2020, 586 (7831): 807-811.

[36] WANG J, ZHU Q, WANG Y, et al. Irisin protects against sepsis-associated encephalopathy by suppressing ferroptosis via activation of the Nrf2/GPX4 signal axis[J]. Free Radical Biology and Medicine, 2022, 187: 171-184.

[37] YU Y Y, JIANG L, WANG H, et al. Hepatic transferrin plays a role in systemic iron homeostasis and liver ferroptosis[J]. Blood, 2020, 136 (6): 726-739.

[38] GAN Z Y, CALLE GARI S, COBBOLD S A, et al. Activation mechanism of PINK1[J]. Nature, 2022, 602 (7896): 328-335.

[39] LEVINE B, KROEMER G. Biological functions of autophagy genes: A disease perspective[J]. Cell, 2019, 176 (1/2): 11-42.

[40] ZOU Y J, XU J J, WANG X, et al. Lactobacillus johnsonii L531 ameliorates Escherichia coli-induced cell damage via inhibiting NLRP3 infl amma some activity and promoting ATG5/ATG16L1-mediated autophagy in porcine mammary epithelial cells[J]. Veterinary Sciences, 2020, 7 (3): 112.

[41] PETHERICK K J, WILLIAMS A C, LANE J D, et al. Autolysosomal β-catenin degradation regulates Wnt-autophagy-p62

crosstalk[J]. The EMBO Journal,2013,32（13）: 1903-1916.
[42] ZHUANG C C, LIU Y, BARK EMA H W, et al. Escherichia coli infection induces ferropto sis in bovine mammary epithelial cells by activating the Wnt/β-catenin pathway-mediated mitophagy[J]. Mitochondrion,2024,78: 101921.

第三章 大肠杆菌型奶牛乳腺炎的防治机制

第一节 番茄红素缓解奶牛乳腺炎的机制研究

一、概述

奶牛乳腺炎通常是由病原体感染引起的,这可能对动物的健康和福利,以及奶牛场的利润和公共卫生产生不利的影响。大肠杆菌是奶牛乳腺炎最常见的病原体之一,主要通过挤奶向外传播,能迅速引起奶牛乳腺炎症,严重时甚至导致奶牛死亡。大肠杆菌是一种可以在动物和人之间传播的病原体,在某些乳制品中被检测到。大肠杆菌感染引起的临床和亚临床乳腺炎受动物管理方式的影响显著。20世纪以来,在处理大肠杆菌感染引起的乳腺炎方面取得了重要进展,但由于种群规模和结构的变化,更严格的加工标准,以及细菌耐药性导致的治疗难题,使得乳腺炎成为乳制品行业的一个复杂而重大的问题。因此,明确大肠杆菌感染引起的牛乳腺炎的致病过程,将为解决细菌耐药问题和探索乳腺炎的治疗方法提供新的思路。

乳体细胞数在正常挤奶过程中迁移到乳中,是世界范围内评价奶牛乳腺健康和奶牛质量的重要指标。乳体细胞是产乳细胞和免疫细胞的混合物。有研究报道,巨噬细胞在正常乳体细胞中所占比例最高(65.53%),而在临床乳腺炎病例中,乳巨噬细胞所占比例明显较低(16.59%),这是由于巨噬细胞在清除致病微生物过程中大量死亡所致。因此,巨噬细胞在病原体引发的牛乳腺炎中发挥了重要作用。小鼠巨噬细胞已成为研究乳腺炎的重要细胞模型。巨噬细胞的不同功能状态极化对于启动对抗细菌和病毒感染的反应至关重要。不同病原体的刺激可以使巨噬细胞转变为经典活化的"M1"或选择性活化的"M2"巨噬细胞，或者处于这两种活化状态之间的某个阶段。M1样巨噬细胞以高诱导型一氧化氮合酶（iNOS）为特征，可促进 TNF-α 和 IL-1β 的产生，增加 M1 标记物 CD86 的表达，保护宿主免受病原体侵害，并有助于抗肿瘤免疫。受周围环境影响，M1 样巨噬细胞可转化为 M2 样巨噬细胞，其特征为 CD206、精氨酸酶 I（ARG1）、几丁质酶 3-like 3（Ym1）和吲哚胺 2,3-双加氧酶（IDO）水平升高，具有抗炎症和介导伤口愈合的作用。大肠杆菌 K88 感染和 LPS/IFN-γ 分别通过 NF-κB 途径激活 M1 样巨噬细胞。然而，从患有严重临床乳腺炎的奶牛乳样中分离出的大肠杆菌菌株对巨噬细胞极化的调节作用尚不清楚。

番茄红素是一种存在于番茄制品、西瓜、葡萄柚和其他水果中的类胡萝卜素，通常用于功能性食品、化妆品和药品中，并因其预防肥胖、糖尿病、癌症、炎症、氧化应激等多种健康问题的能力而受到越来越多的关注。番茄红素不仅可以通过 NFE2L2 信号通路减少 H_2O_2 引起的各种病理变化如炎症，还可以通过介导巨噬细胞 M1/M2 状态来缓解肥胖引起的炎症。但番茄红素是否通过巨噬细胞极化减轻大肠杆菌感染引起的炎症反应尚不清楚。因此，我们利用大肠杆菌感染巨噬细胞的实验模型，研究番茄红素是否可以通过调节巨噬细胞极化来提供抗炎保护，这为鼓励寻找除抗生素以外的治疗牛乳腺炎的方法提供了新的见解。

二、不同浓度番茄红素对巨噬细胞的细胞毒性作用

与未处理的巨噬细胞相比，0.5 μM 番茄红素显著抑制了 LDH 活性（$P<0.05$，图 3-1A）且巨噬细胞存活率显著升高（$P<0.01$，图 3-1B）；而 2 μM 和 3 μM 番茄红素可显著增加 LDH 活性（$P<0.01$），3 μM 番茄红素可显著降低巨噬细胞活力（$P<0.05$）。

三、番茄红素抑制大肠杆菌感染的巨噬细胞的 LDH 活性增加

与未处理的巨噬细胞相比，0.5 μM 番茄红素处理抑制了 LDH 的释放（$P<0.05$，图 3-1C），大肠杆菌感染显著增加 LDH 活性（$P<0.01$，图 3-1C）。与大肠杆菌感染的巨噬细胞相比，番茄红素处理（$P<0.01$，图 3-1C）显著抑制 LDH 释放量（$P<0.01$，图 3-1C）。

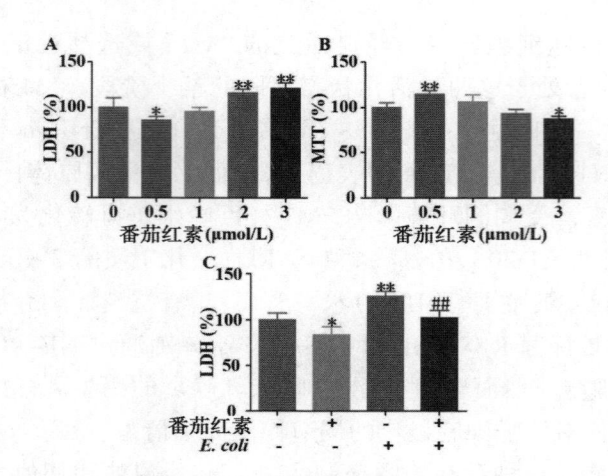

图 3-1　番茄红素处理和/或大肠杆菌感染对巨噬细胞的影响

四、番茄红素处理抑制大肠杆菌感染诱导的 M1 样巨噬细胞极化

与未处理巨噬细胞相比,0.5 μM 番茄红素处理显著抑制巨噬细胞内 iNOS、CD86、TNF-α 和 IL-1β 蛋白表达水平（$P < 0.05$；$P < 0.01$，图 3-2）。大肠杆菌感染显著增加 iNOS、CD86、TNF-α 和 IL-β 蛋白表达水平，TNF-α 和 IL-1β 含量升高（$P < 0.01$，图 3-2）。与大肠杆菌感染的巨噬细胞相比，番茄红素显著抑制了 iNOS、CD86、TNF-α 和 IL-β 蛋白水平以及 TNF-α 和 IL-1β 含量（$P < 0.01$）。

五、番茄红素处理促进大肠杆菌感染诱导的 M2 样巨噬细胞极化

与未处理的巨噬细胞相比，0.5 μM 番茄红素处理显著增加巨噬细胞 CD206、ARG1、IDO 和 YM1 蛋白表达水平（$P < 0.01$）。大肠杆菌感染显著降低巨噬细胞 CD206、ARG1、IDO 和 YM1 蛋白表达水平（$P < 0.01$，图 3-3）。与大肠杆菌感染的巨噬细胞相比，番茄红素处理显著增加了 CD206、ARG1、IDO 和 YM1 蛋白表达水平（$P < 0.01$，图 3-3）。

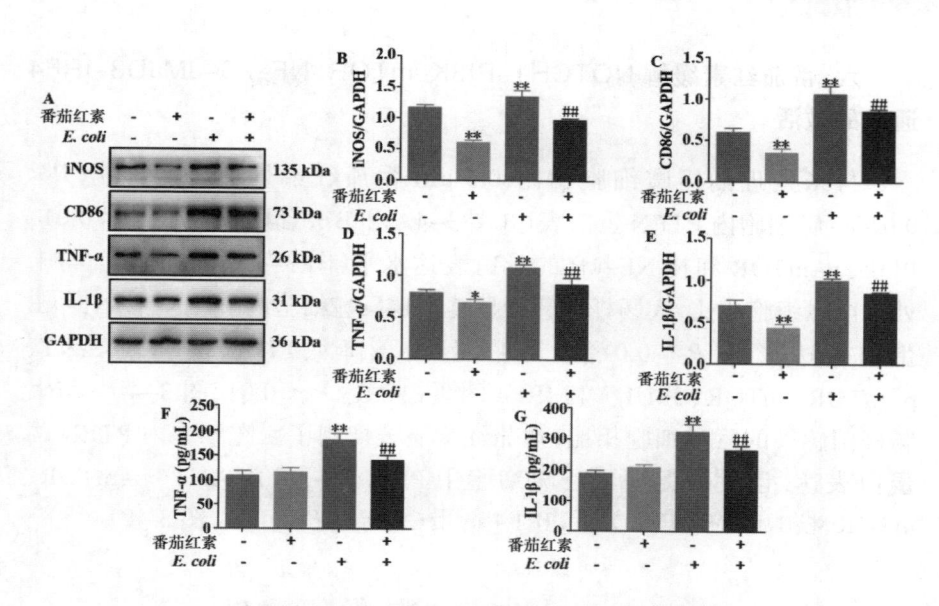

图 3-2 番茄红素减少大肠杆菌感染，增加 M1 样巨噬细胞

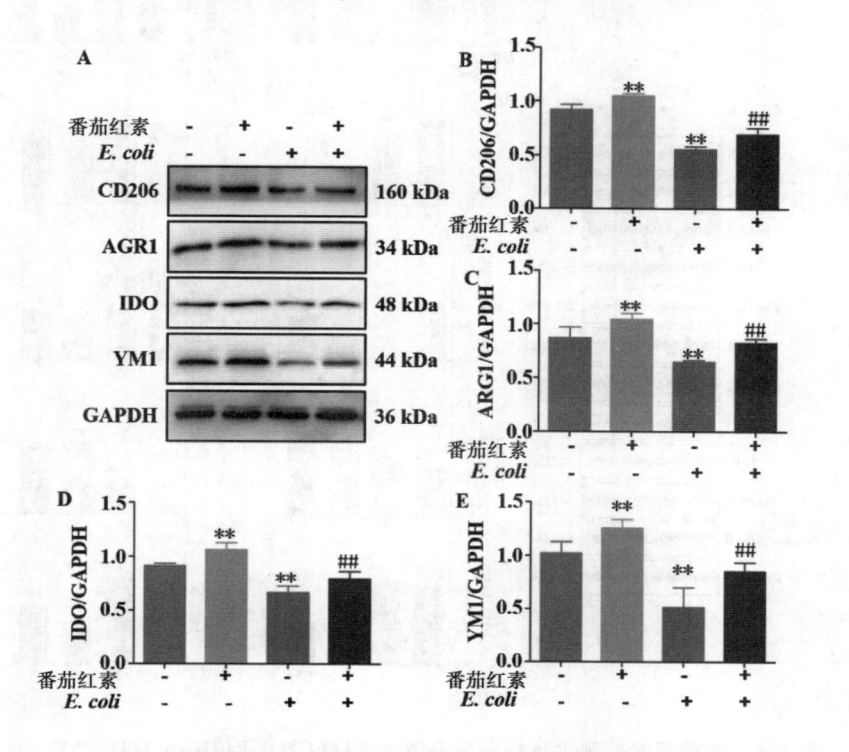

图 3-3 番茄红素减少大肠杆菌感染抑制 M2 样巨噬细胞

六、番茄红素缓解 NOTCH1-PI3K-mTOR-NF-κB-JMJD3-IRF4 通路的激活

与未处理的巨噬细胞相比，0.5 μM 番茄红素处理显著抑制（$P < 0.01$）巨噬细胞内 PTEN 蛋白表达（图 3-4A 和 B），且显著增加（$P < 0.01$）PI3k、p-mTOR 和核 NF-κB 的蛋白表达水平（图 3-4A、F 和 H）。与未处理巨噬细胞相比，大肠杆菌感染显著增加巨噬细胞内 PTEN、核 NF-κB 蛋白表达水平（$P < 0.01$，图 3-4），且显著降低 PI3K、p-AKT、AKT、p-mTOR、mTOR、JMJD3 和 IRF4 的蛋白表达 $P < 0.01$，图 3-4）。与大肠杆菌感染的巨噬细胞相比，番茄红素显著抑制了巨噬细胞内 PTEN 的蛋白表达，且显著促进了巨噬细胞中 PI3K、p-AKT、AKT、p-mTOR、mTOR、核 NF-κB、JMJD3 和 IRF4 的蛋白表达（$P < 0.01$，图 3-4）。

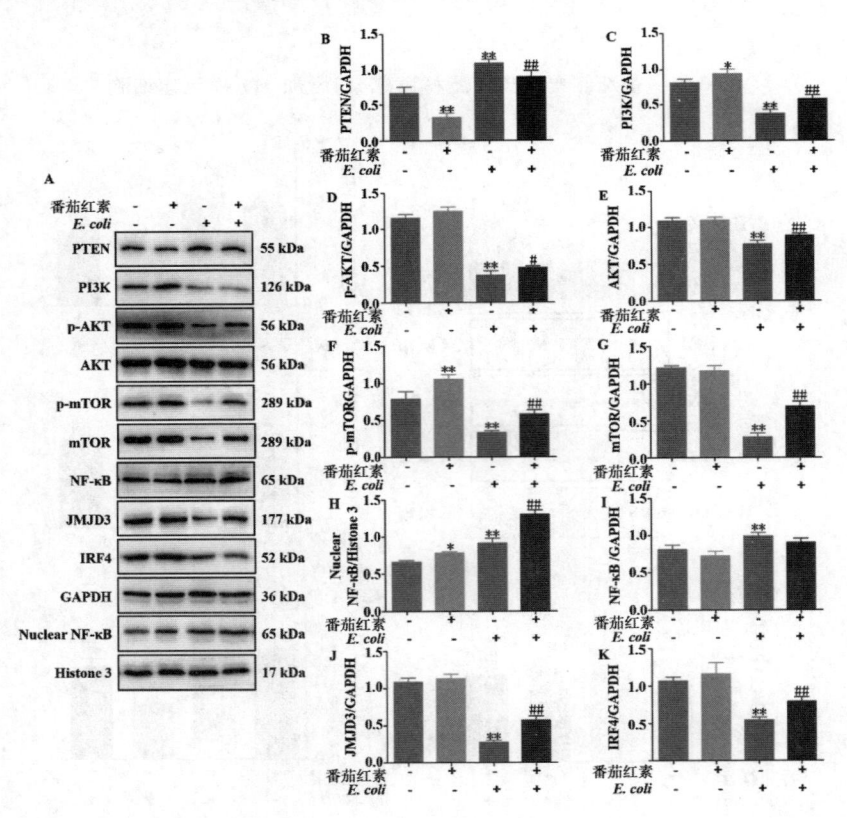

图 3-4 番茄红素抑制大肠杆菌感染激活 NOTCH1-PI3K-mTOR-NF-κB-JMJD3-IRF4 通路

七、NOTCH1-PI3K-mTOR-NF-κB-JMJD3-IRF4 通路在 M2 样巨噬细胞极化中的作用

为了阐明 NOTCH1-PI3K-mTOR-NF-κB-JMJD3-IRF4 通路在番茄红素缓解大肠杆菌感染巨噬细胞中的作用，使用了 PTEN 抑制剂 Ginkgolic acid C17∶1、PI3K 激活剂 740YPDGFR、AKT 激活剂 SC79 和 NF-κB 激活剂 CHPG 钠盐来抑制该通路。与番茄红素和大肠杆菌共同处理组相比，Ginkgolic acid C17∶1、740YPDGFR、SC79 和 CHPG 钠盐分别抑制了 CD206、ARG1、IDO 和 YM1 的蛋白表达（$P < 0.01$，图 3-5）。

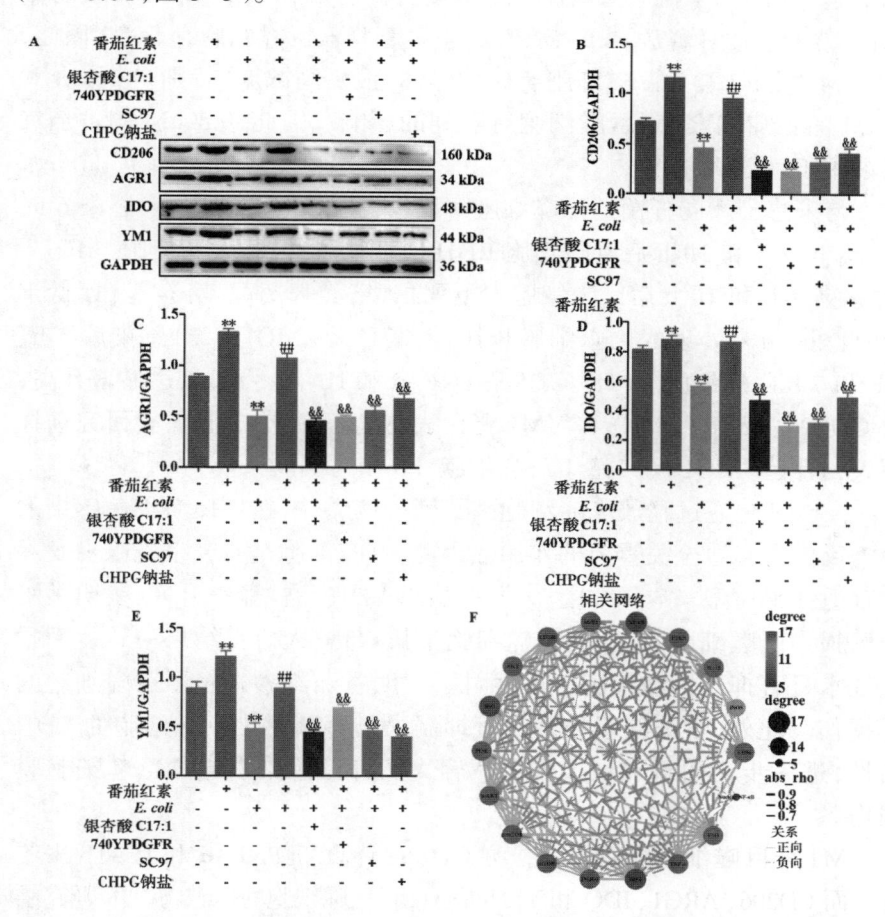

图 3-5 NOTCH1-PI3K-mTOR-NF-κB-JMJD3-IRF4 通路在大肠杆菌感染巨噬细胞中番茄红素促进的 M2 样巨噬细胞极化中的作用

八、讨论

本研究的目的是确定番茄红素是否能通过介导巨噬细胞极化减轻大肠杆菌感染诱发的炎症反应。在研究中,前期分离得到的大肠杆菌被用于感染小鼠巨噬细胞 J774A.1 细胞系。本研究发现大肠杆菌感染通过激活 NOTCH1-PI3K-mTOR-NF-κB-JMJD3-IRF4 通路诱导 M1 样巨噬细胞极化,而番茄红素缓解了该作用。

在本研究中,0.5 μM 番茄红素能显著抑制巨噬细胞 LDH 活性并提高细胞存活率。然而,当番茄红素浓度升高至 2 μM 和 3 μM 却显著增加巨噬细胞 LDH 释放,3 μM 番茄红素显著降低巨噬细胞存活率。因此,0.5 μM 番茄红素对巨噬细胞有保护作用,但更高浓度(特别是 2 μM 和 3 μM 的番茄红素)对巨噬细胞有不利的影响。因此,0.5 μM 的番茄红素被用于后续研究。

与未感染大肠杆菌的巨噬细胞相比,大肠杆菌(MOI = 5)感染 6 h、7 h、8 h、9 h 和 10 h 后巨噬细胞 LDH 活性显著升高,大肠杆菌(MOI = 5)感染 8 h 和 10 h 后巨噬细胞大量死亡,培养基变黄,培养基 pH 低于 6。此外,与未处理的巨噬细胞相比,大肠杆菌(MOI = 5)感染后 5 h、6 h 和 7 h 巨噬细胞 iNOS、CD86、TNF-α 和 IL-1β 蛋白表达显著升高,而 CD206、AGR1、IDO 和 YM1 蛋白表达显著降低。因此,本研究选择用大肠杆菌(MOI = 5)感染巨噬细胞 6 h 作为实验条件。

巨噬细胞通过释放炎症细胞因子消灭病原体,在机体对细菌感染的免疫反应中起着至关重要的作用。因此,巨噬细胞是研究牛乳腺炎发展的合适细胞模型。本研究中,大肠杆菌组 LDH 活性显著升高,表明大肠杆菌损伤巨噬细胞膜,抑制巨噬细胞生长,与前人研究结果一致。而番茄红素显著抑制巨噬细胞 LDH 活性,表明番茄红素减轻了巨噬细胞的膜损伤。此外,番茄红素处理显著抑制大肠杆菌感染巨噬细胞的 LDH 活性,进一步证明番茄红素可以减轻大肠杆菌感染引起的巨噬细胞膜损伤。

M1 样巨噬细胞可以通过检测 CD86、iNOS、TNF-α 和 IL-1β 来鉴定。而 CD206、ARG1、IDO 和 YM1 是 M2 样巨噬细胞的标志物。本研究结果显示,大肠杆菌感染显著升高 iNOS、CD86、TNF-α 和 IL-1β 的表达,而番茄红素显著抑制了这些蛋白的表达。因此,大肠杆菌感染诱导 M1

样巨噬细胞，而番茄红素减少 M1 样巨噬细胞。此外，番茄红素显著降低了 CD206、ARG1、IDO 和 YM1 蛋白水平。因此，大肠杆菌感染抑制 M2 样巨噬细胞的产生，而番茄红素促进 M2 样巨噬细胞的产生。此外，延长大肠杆菌感染时间可显著升高 iNOS、CD86、TNF-α 和 IL-1β 蛋白表达，抑制 CD206、AGR1、IDO 和 YM1 蛋白水平。可见，大肠杆菌感染对巨噬细胞极化的影响具有时间依赖性。此外，番茄红素处理显著降低了大肠杆菌感染巨噬细胞中 iNOS、CD86、TNF-α 和 IL-1β 表达的升高，缓解了 CD206、ARG1、IDO 和 YM1 蛋白水平的下降。这些结果表明，番茄红素能够通过控制巨噬细胞极化来减少炎症反应。

Notch 信号通路主要通过抑制 PTEN 参与细胞的分化、增殖、凋亡和炎症过程。PTEN 通过抑制 PI3K、AKT 和 mTOR 信号通路调节各种重要的细胞活动（如生长、生存和代谢）。PI3K-AKT-mTOR 信号通路可以介导细胞极性和膜重塑。PI3K 可被生长因子刺激激活，并通过产生磷脂酰肌醇 3,4,5-二磷酸结合 AKT。AKT 的完全激活需要 Ser473 和 Thr308 的磷酸化。同时，Ser2448 位点发生磷酸化可被用于评估 mTOR 激酶的激活状态或作为 mTOR 途径启动的间接测量指标。激活 mTOR 通过促进 NF-κB 在细胞核中的释放和易位来促进促炎细胞因子和 JMJD3 在微生物感染反应中的表达。JMJD3 负责去除被抑制的三甲基化组蛋白标记，以增强靶基因的表达，特别是 IRF4。JMJD3-IRF4 通路也参与了巨噬细胞的活化和极化。大肠杆菌显著促进 PTEN、NF-κB 和核 NF-κB 蛋白表达，降低 PI3K、p-AKT、AKT、p-mTOR、mTOR、JMJD3 和 IRF4 蛋白表达，提示本研究中大肠杆菌感染激活了 NOTCH1-PI3K-mTOR-NF-κB-JMJD3-IRF4 通路。而番茄红素显著降低大肠杆菌感染巨噬细胞中 PTEN 蛋白水平，降低 PI3K、p-AKT、AKT、p-mTOR、mTOR、JMJD3 和 IRF4 蛋白水平。因此，番茄红素在大肠杆菌感染的巨噬细胞中抑制 NOTCH1-PI3K-mTOR-NF-κB-JMJD3-IRF4 通路的激活。此外，PTEN 抑制剂 Ginkgolic acid C17∶1、PI3K 活化剂 740YPDGFR、AKT 激活剂 SC79 和 NF-κB 激活剂 CHPG 钠盐，在番茄红素和大肠杆菌共同处理的巨噬细胞中，分别显著地降低 CD206、AGR1、IDO 和 YM1 蛋白表达。因此，番茄红素通过抑制 NOTCH1-PI3K-mTOR-NF-κB-JMJD3-IRF4 通路介导巨噬细胞极化，从而减轻大肠杆菌感染诱导的牛乳腺炎。

九、结论

番茄红素在巨噬细胞中通过抑制 NOTCH1-PI3K-mTOR-NF-κB-JMJD3-IRF4 通路抑制 M1 样巨噬细胞极化并增加 M2 样巨噬细胞以应对大肠杆菌感染（图 3-6）。本研究将为番茄红素的细胞保护作用提供实验依据，为预防和治疗大肠杆菌感染引起的牛乳腺炎提供新的实验知识。因此，番茄红素可能是治疗牛乳腺炎的潜在药物，但其药效学研究需要进一步在乳腺炎动物模型中进行评估。

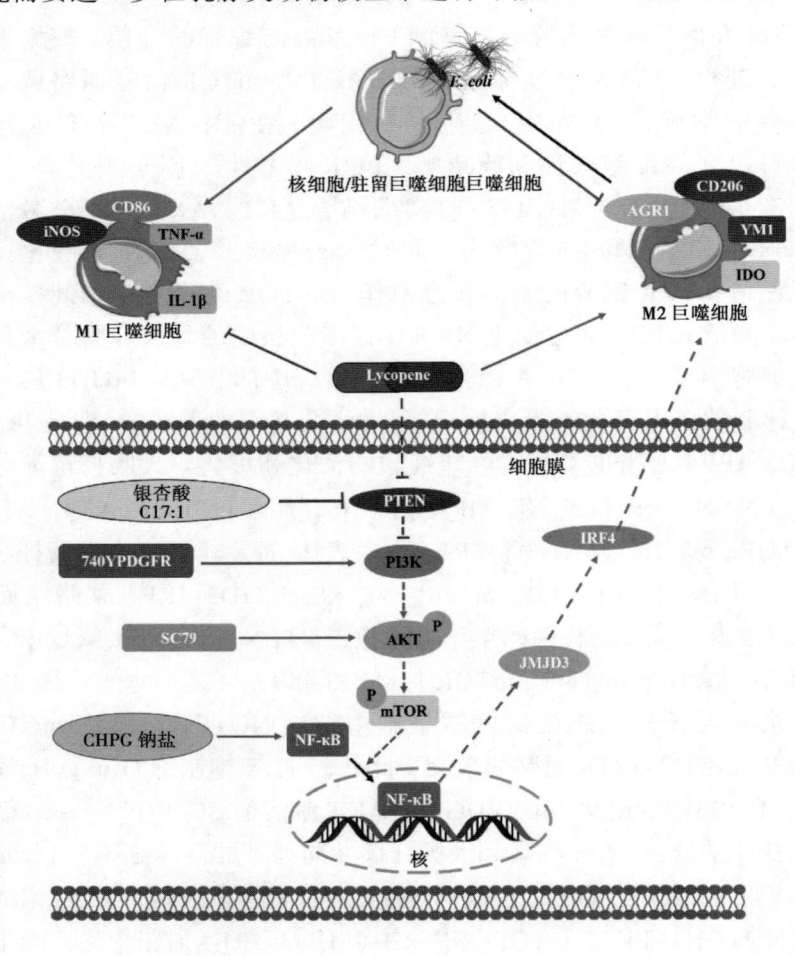

图 3-6 番茄红素对大肠杆菌感染巨噬细胞的保护作用

参考文献

[1] SHARUN K, DHAMA K, TIWARI R, et al. Advances in therapeutic and managemental approaches of bovine mastitis: A comprehensive review[J]. The Veterinary Quarterly, 2021, 41（1）: 107-136.

[2] GAO J, BARKEMA H W, ZHANG L M, et al. Incidence of clinical mastitis and distribution of pathogens on large Chinese dairy farms[J]. Journal of Dairy Science, 2017, 100（6）: 4797-4806.

[3] CVETNIC L, SAMARDZIJA M, DUVNJAK S, et al. Multi locus sequence typing and spa typing of Staphylococcus aureus isolated from the milk of cows with subclinical mastitis in Croatia[J]. Microorganisms, 2021, 9（4）: 725.

[4] LI Y N, ZHU Y H, CHU B X, et al. Map of enteropathogenic Escherichia coli targets mitochondria and triggers DRP-1-mediated mitochondrial fission and cell apoptosis in bovine mastitis[J]. International Journal of Molecular Sciences, 2022, 2（9）: 4907.

[5] GARCIA A. Contagious vs. environmental mastitis[J]. Extension Extra, 2004, 126: 1-2.

[6] RUEGG P L. A 100-Year Review: Mastitis detection, management, and prevention[J]. Journal of Dairy Science, 2017, 100（12）: 10381-10397.

[7] ALHUSSIEN M N, DANG A K. Milk somatic cells, factors influencing their release, future prospects, and practical utility in dairy animals: An overview[J]. Veterinary World, 2018, 11（5）: 562-577.

[8] ALHUSSIEN M, MANJARI P, MOHAMMED S, et al. Incidence of mastitis and activity of milk neutrophils in Tharparkar cows reared under semi-arid conditions[J]. Tropical Animal Health and Production, 2016, 48（6）: 1291-1295.

[9] ZHAO C J, BAO L J, ZHAO Y H. et al. A fiber-enriched diet alleviates Staphylococcus aureus-induced mastitis by activating the HDAC3-mediated antimicrobial program in macrophages via

butyrate production in mice[J]. PLoS Pathogens, 2023, 19 (1): e1011108.

[10] SWARTZ T H, BRADFORD B J, MAME DO VA L K. Connecting metabolism to mastitis: Hyperketonemia impaired mammary gland defenses during a Streptococcus uberis challenge in dairy cattle[J]. Frontiers in Immunology, 2021, 12: 700278.

[11] ZHANG W, FU Z D, YIN H Y, et al. Macrophage polarization modulated by porcine circovirus type 2 facilitates bacterial coinfection[J]. Frontiers in Immunology, 2021, 12: 688294.

[12] HOWARD N C, KHADER S A. Immunometabolism during Mycobacterium tuberculosis Infection[J]. Trends in Microbiology, 2020, 28 (10): 832-850.

[13] GE G, JIANG H Q, XIONG J S, et al. Progress of the art of macrophage polarization and different subtypes in mycobacterial infection[J]. Frontiers in Immunology, 2021, 12: 752657.

[14] LUMENG C N, BODZIN J L, SALTIEL A R. Obesity induces a phenotypic switch in adipose tissue macrophage polarization[J]. The Journal of Clinical Investigation, 2007, 117 (1): 175-184.

[15] WANG X F, WANG H S, WANG H, et al. The role of indoleamine 2,3-dioxygenase (IDO) in immune tolerance: Focus on macrophage polarization of THP-1 cells[J]. Cellular Immunology, 2014, 289 (1/2): 42-48.

[16] CHEN S, LU Z Q, WANG F Q, et al. Cathelicidin-WA polarizes E.coli K88-induced M1 macrophage to M2-like macrophage in $RAW_{264.7}$ cells[J]. International Immunopharmacology, 2018, 54: 52-59.

[17] HUANG Y L, TIAN C, LI Q M, et al. TET1 knockdown inhibits Porphyromonas gingivalis LPS/IFN-γ-induced M1 macrophage polarization through the NF-κB pathway in THP-1 cells[J]. International Journal of Molecular Sciences, 2019, 20 (8): 2023.

[18] ZHU R Y, CHEN B B, BAI Y, et al. Lycopene in protection against obesity and diabetes: A mechanistic review[J]. Pharmacological Research, 2020, 159: 104966.

[19] SUN X D, JIA H D, XU Q S, et al. Lycopene alleviates H_2O_2-induced oxidative stress, inflammation and apoptosis in bovine mammary epithelial cells via the NFE2L2 signaling pathway[J]. Food and Function, 2019, 10（10）: 6276-6285.

[20] CHEN G L, NI Y H, NAGATA N, et al. Lycopene alleviates obesity-induced inflammation and insulin resistance by regulating M1/ M2 status of macrophages[J]. Molecular Nutrition and Food Research, 2019, 63（21）: e1900602.

[21] ZHUANG C C, ZHAO J H, ZHANG S Y, et al.Escherichia coli infection mediates pyroptosis via activating p53-p21 pathway-regulated apoptosis and cell cycle arrest in bovine mammary epithelial cells[J]. Microbial Pathogenesis, 2023, 184: 106338.

[22] RAFI M M, YADAV P N, REYES M. Lycopene inhibits LPS-induced proinflammatory mediator inducible nitric oxide synthase in mouse macrophage cells[J]. Journal of Food Science, 2007, 72（1）: 69-74.

[23] GALLI G, SALEH M. Immunometabolism of macrophages in bacterial infections[J]. Frontiers in Cellular and Infection Microbiology, 2021, 10: 607650.

[24] DUCHESNE C, FRESCALINE N, BLAISE O, et al.Cold atmospheric plasma promotes killing of Staphylococcus aureus by macrophages[J]. mSphere, 2021, 6（3）: e0021721.

[25] LIMA E S, BLAGITZ M G, BATISTA C F, et al. Milk macrophage function in bovine leukemia virus-infected dairy cows[J]. Frontiers in Veterinary Science, 2021, 8: 650021.

[26] ZHUANG C C, LIU G, BARKEMA H W, et al. Selenomethionine suppressed TLR4/NF-κB pathway by activating selenoprotein S to alleviate ESBL Escherichia coli-induced inflammation in bovine mammary epithelial cells and macrophages[J]. Frontiers in Microbiolgy, 2020, 11: 1461.

[27] FANG J Q, OU Q, WU B H, et al. TcpC inhibits M1 but promotes M2 macrophage polarization via regulation of the MAPK/NF-κB and akt/ STAT6 pathways in urinary tract infection[J]. Cells,

2022,11(17):2674.

[28] CUTOLO M, CAMPITIELLO R, GOTELLI E, et al. The role of M1/ M2 macrophage polarization in rheumatoid arthritis synovitis[J]. Frontiers in Immunology,2022,13:867260.

[29] LIN Y W, LI X X, FU F H, et al. Notch1/Hes1-PTEN/AKT/ IL-17A feedback loop regulates Th17 cell differentiation in mouse psoriasis-like skin inflammation[J]. Molecular Medicine Reports, 2022,26(1):223.

[30] ÁLVAREZ-GARCIA V, TAWIL Y, WISE H M, et al. Mechanisms of PTEN loss in cancer: It's all about diversity[J]. Seminars in Cancer Biology,2019,59:66-79.

[31] KRIPLANI N, HERMIDA M A, BROWN E R, et al. Class I PI3-kinases: Function and evolution[J]. Advances in Biological Regulation,2015,59:53-64.

[32] WANG Y F, KURAM ITSU Y, BARON B, et al. PI3K inhibitor LY294002, as opposed to wortmannin, enhances AKT phosphorylation in gemcitabine-resistant pancreatic cancer cells[J]. International Journal of Oncology,2017,50(2):606-612.

[33] KIZHAKKAYIL J, THAYYULLATHIL F, CHATHOTH S, et al. Modulation of curcumin-induced Akt phosphorylation and apoptosis by PI3K inhibitor in MCF-7 cells[J]. Biochemical and Biophysical Research Communications,2010,394(3):476-81.

[34] FIGUEIREDO V C, MARKWORTH J F, CAMERON-SMITH D. Considerations on mTOR regulation at serine 2448: Implications for muscle metabolism studies[J]. Cellular and Molecular Life Sciences,2017,74(14):2537-2545.

[35] DE SANTA F, NARANG V, YAP Z H, et al. Jmjd3 contributes to the control of gene expression in LPS-activated macrophages[J]. The EMBO Journal,2009,28(21):3341-3352.

[36] ZHOU M X, XU W M, WANG J Z, et al. Boosting mTOR-dependent autophagy via upstream TLR4-MyD88-MAPK signalling and downstream NF-κB pathway quenches intestinal

inflammation and oxidative stress injury[J]. EBioMedicine, 2018, 35: 345-360.

[37] MING-CHIN LEE K, ACHUTHAN A A, DE SOUZA D P, et al. Type I interferon antagonism of the JMJD3-IRF4 pathway modulates macrophage activation and polarization[J]. Cell Reports, 2022, 39 (3): 110719.

[38] SATOH T, TAKEUCHI O, VANDENBON A, et al. The Jmjd3-Irf4 axis regulates M2 macrophage polarization and host responses against helminth infection[J]. Nature Immunology, 2010, 11(10): 936-944.

第二节 硒缓解大肠杆菌感染致炎症的机制研究

一、概述

大肠杆菌是一种影响奶牛乳腺的机会性环境病原菌,可引起牛临床乳腺炎,特别是在奶牛哺乳期早期。大肠杆菌通过 LDH 的释放进入乳腺腔内,然后引起乳腺应对感染出现的炎症和先天免疫反应。当感染发生时,乳腺上皮细胞和巨噬细胞是抵御侵袭性乳腺炎病原体的第一道防线。因此,标准、便宜、容易获得的奶牛乳腺上皮细胞和巨噬细胞是评估大肠杆菌感染诱发炎症的理想体外试验模型。此外,乳腺上皮细胞和巨噬细胞拥有一系列种系编码的模式识别受体/传感器,这些受体/传感器能识别病原体相关的分子模式,并激活 NF-κB 信号通路,介导炎症和先天免疫反应。在细胞受到刺激后,与细胞质中的 IκB 蛋白结合的 NF-κB 被释放并易位到细胞核中,NF-κB 的 p65 亚基含有转录激活结构域,在感染过程中调控炎症蛋白的转录如 TNF-α 和 IL-1β。

硒是人体必需的微量元素,富硒饮食是哺乳动物硒摄入的主要来源。硒的主要代表化学形式是无机硒(亚硒酸盐和硒酸盐)和有机硒(硒半胱氨酸、硒代蛋氨酸、甲基硒半胱氨酸和富硒酵母),这些硒形式参与多种生化和生理功能,如免疫、抗氧化和抗衰老。此外,硒可以调节炎症过程,硒代合物可以影响巨噬细胞功能。此外,添加硒酵母的饮食增加了血液中性粒细胞杀死大肠杆菌的百分比。然而,硒代蛋氨酸对大肠杆菌感染奶牛乳腺上皮细胞和巨噬细胞的作用及其分子机制尚不清楚。

二、硒代蛋氨酸对奶牛乳腺上皮细胞和巨噬细胞的细胞毒作用

为了评估硒的细胞毒作用(图 3-7),不同浓度的硒代蛋氨酸被用于培养奶牛乳腺上皮细胞和巨噬细胞 12 h,并测量细胞活力和 LDH 活性。与空白对照组相比较,40 μM 硒代蛋氨酸显著提高奶牛乳腺上皮细胞存

活率（$P < 0.01$，图 3-7A）并显著抑制 LDH 活性（$P < 0.01$，图 3-7B）。此外，与空白对照组相比，40 μM 和 60 μM 硒代蛋氨酸增加了巨噬细胞活力（图 3-7C），且显著降低了 LDH 释放（图 3-7D），表明这些浓度下硒代蛋氨酸无细胞毒性（$P < 0.01$）。与空白对照组相比较，80 μM 和 100 μM 硒代蛋氨酸显著抑制奶牛乳腺上皮细胞和巨噬细胞的细胞活力，显著提高 LDH 活性（$P < 0.01$，图 3-7A 和 B），表明这些浓度下硒代蛋氨酸有细胞毒性。

三、硒代蛋氨酸降低奶牛乳腺上皮细胞和巨噬细胞内 LDH 活性

与空白对照组相比，大肠杆菌感染显著增加奶牛乳腺上皮细胞和巨噬细胞中 LDH 活性（$P < 0.01$，图 3-7E 和 F），5～40 μM 硒代蛋氨酸在奶牛乳腺上皮细胞中显著缓解了该作用（$P < 0.05$，图 3-7E），20～60 μM 硒代蛋氨酸在巨噬细胞中显著缓解了该作用（$P < 0.05$，图 3-7F）。

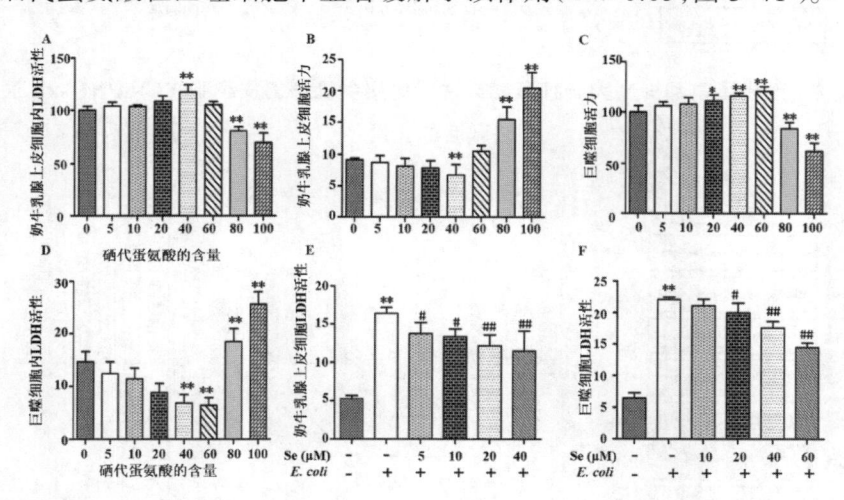

图 3-7 硒代蛋氨酸和/或大肠杆菌感染对奶牛乳腺上皮细胞和巨噬细胞的影响

四、硒代蛋氨酸可增加 SeS 并抑制 TLR4 介导的 NF-κB 信号通路

与大肠杆菌感染组相比，5～40 μM 硒代蛋氨酸显著增加了大肠杆菌诱导的奶牛乳腺上皮细胞 SeS 蛋白表达的下调（$P < 0.05$，图 3-8A 和 B），且 20～60 μM 硒代蛋氨酸显著增加了大肠杆菌诱导的巨噬细

胞 SeS 蛋白表达的下调（图 3-9A 和 B）。与对照组相比，大肠杆菌感染显著增加（$P<0.01$）了 TLR4、IκB、TNF-α 和 IL-1β 蛋白表达，以及 NF-κB 核蛋白表达（图 3-8A、E，图 3-9A、E，$P<0.01$），而硒代蛋氨酸预处理可缓解该作用（图 3-8，图 3-9，$P<0.01$）。

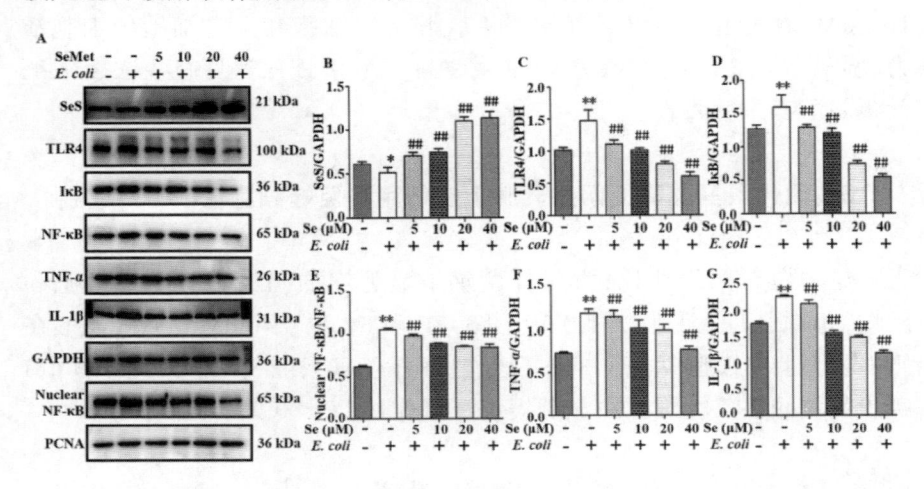

图 3-8 硒代蛋氨酸抑制大肠杆菌感染诱导的奶牛乳腺上皮细胞 TLR4/NF-κB 通路激活的作用

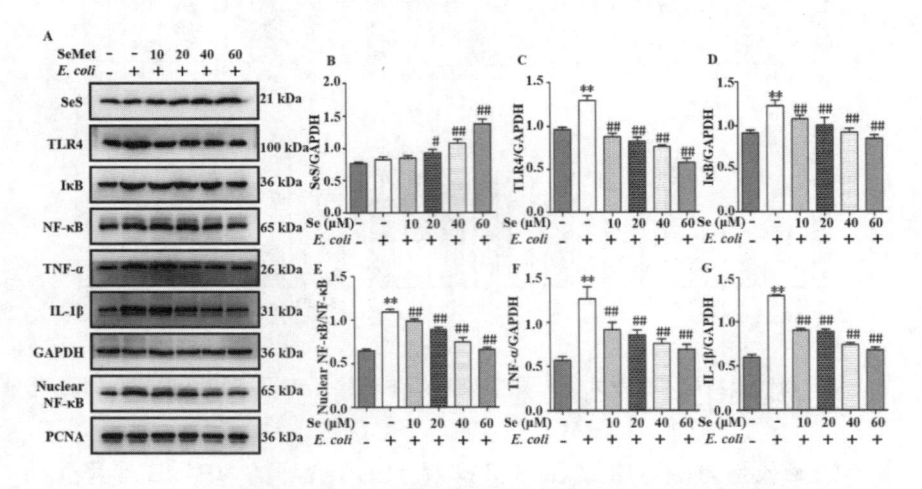

图 3-9 硒代蛋氨酸抑制大肠杆菌感染诱导的巨噬细胞 TLR4/NF-κB 通路激活的作用

五、大肠杆菌感染可促进 MCP-1、CCL-3 和 CCL-5 mRNA 的表达

与对照组相比,大肠杆菌感染显著增加了奶牛乳腺上皮细胞中 MCP-1、CCL-3 和 CCL-5 的 mRNA 表达($P < 0.01$,图 3-10)。

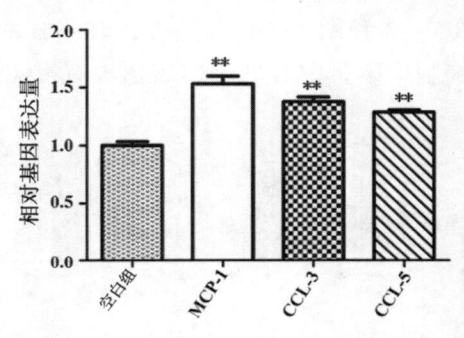

图 3-10　大肠杆菌感染对奶牛乳腺上皮细胞 MCP-1、CCL-3 和 CCL-5 mRNA 表达的影响

六、硒蛋白 S 缺失可逆转硒代蛋氨酸对上皮细胞和巨噬细胞的作用

与硒代蛋氨酸和大肠杆菌共同处理组相比,SeS 敲低显著缓解($P < 0.01$,$P < 0.05$)TLR4、IKB、TNF-α 和 IL-1β 蛋白表达及 NF-κB 核蛋白表达的降低。

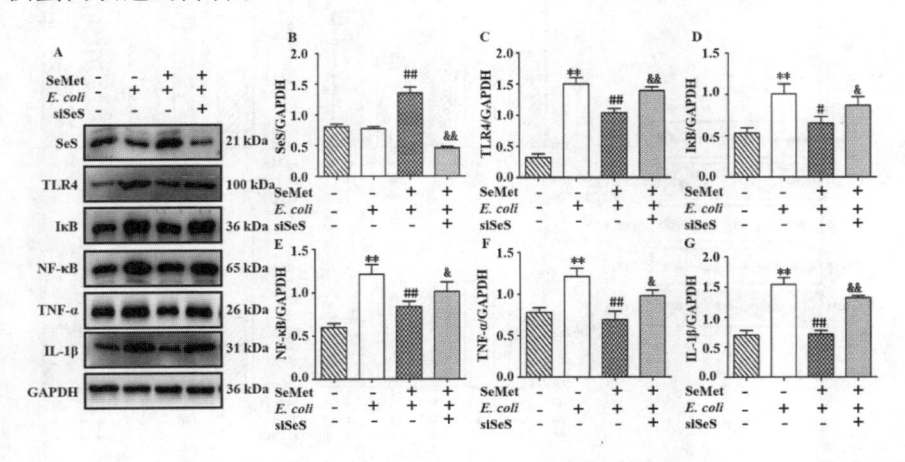

图 3-11　硒蛋白 S 缺失逆转硒代蛋氨酸对大肠杆菌感染处理的奶牛乳腺上皮细胞 TLR4/NF-κB 通路的影响

七、NF-κB 参与硒代蛋氨酸缓解炎症

与大肠杆菌感染组相比较，NF-κB 抑制剂 BAY 11-708 在奶牛乳腺上皮细胞和巨噬细胞中，显著抑制了（$P < 0.05$）TNF-α 和 IL-1β 的蛋白表达水平（图 3-13 和图 3-14）。此外，NF-κB 的降低对奶牛乳腺上皮细胞和巨噬细胞中硒蛋白 S 的蛋白表达没有影响。

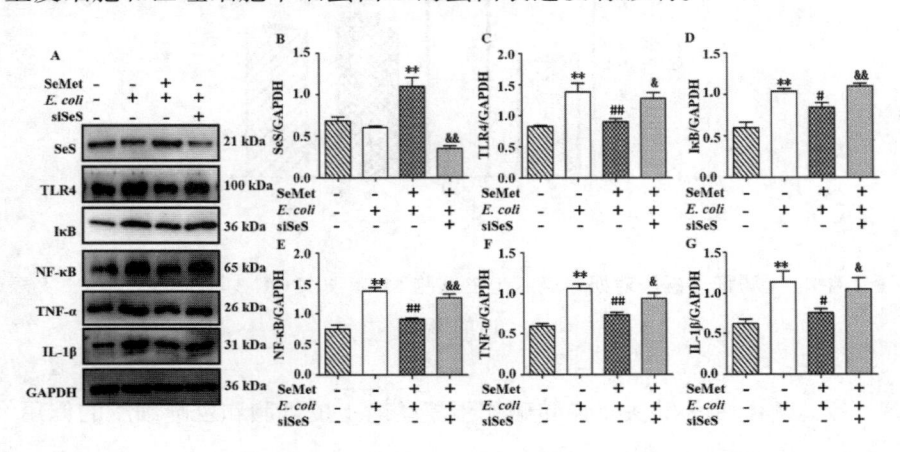

图 3-12　硒蛋白 S 缺失逆转硒代蛋氨酸对大肠杆菌感染处理的巨噬细胞 TLR4/NF-κB 通路的影响

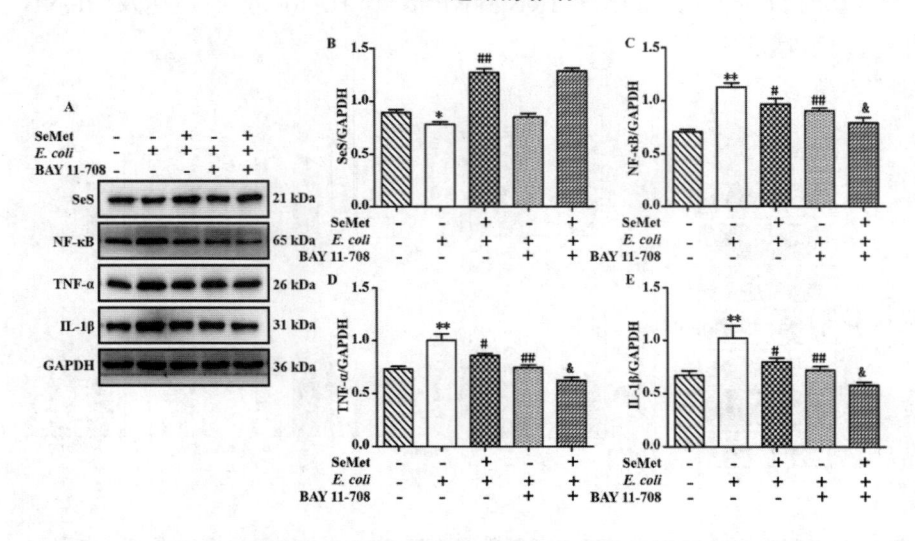

图 3-13　NF-κB 参与硒代蛋氨酸对大肠杆菌感染诱导的奶牛乳腺上皮细胞炎症的调控

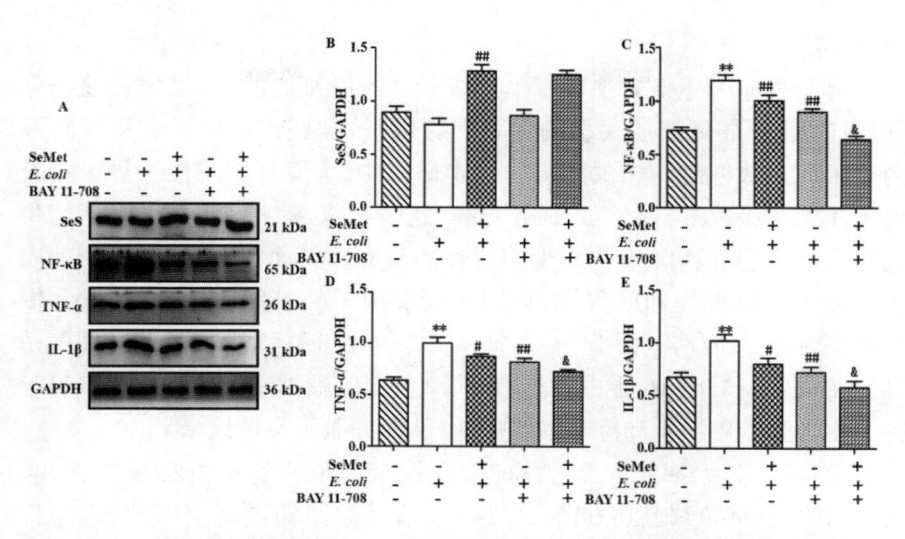

图3-14　NF-κB参与硒代蛋氨酸对大肠杆菌感染诱导的巨噬细胞炎症的调控

八、讨论

在本研究中,40 μM和/或60 μM硒代蛋氨酸分别可显著提高奶牛乳腺上皮细胞和巨噬细胞的细胞活力并抑制LDH释放,而80 μM和100 μM硒代蛋氨酸则损伤了奶牛乳腺上皮细胞和巨噬细胞,这表明硒代蛋氨酸在0～40 μM浓度范围对奶牛乳腺上皮细胞、在0～60 μM浓度范围对巨噬细胞具有保护作用,80 μM和100 μM硒代蛋氨酸对奶牛乳腺上皮细胞和巨噬细胞具有毒性作用,与以往研究结果一致。此外,研究报道4 μM、8 μM、16 μM、32 μM、64 μM和128 μM亚硒酸钠在12 h时显著降低了$RAW_{264.7}$巨噬细胞的细胞活力。预处理10 μM亚硒酸蛋氨酸或1 μM亚硒酸钠能显著缓解奶牛乳腺上皮细胞的细胞活力(分别为83.1%和81.4%)。这种差异可能是由于不同形式的硒在吸收机制、生物利用度和毒性方面存在差异,且有机硒比无机硒具有更高的生物利用度,对动物的毒性较小,这与其他研究结果一致。在本研究中,预补充硒代蛋氨酸可降低奶牛乳腺上皮细胞和巨噬细胞中大肠杆菌感染诱导的LDH活性的增加。LDH作为一种糖酵解酶,在体内大多数类型的细胞中含量丰富,LDH释放量增加表明细胞膜通透性增加,细胞膜破裂。因此,大肠杆菌感染损伤了细胞膜完整性,而硒代蛋氨酸则减轻了大肠杆菌感染诱导的奶牛乳腺上皮细胞和巨噬细胞细胞膜的损伤。

当大肠杆菌感染奶牛乳腺上皮细胞和巨噬细胞时,会引发快速的炎症反应,其特征是释放大量的促炎细胞因子,这些细胞因子在第一阶段消除了免疫效应细胞中的感染。在本研究中,大肠杆菌感染诱导了促炎因子 TNF-α 和 IL-1β 含量显著升高,这表明大肠杆菌感染引起了奶牛乳腺上皮细胞和巨噬细胞发生剧烈炎症反应。然而,促炎细胞因子的表达必须被严格调控和区隔化,以防止促炎因子的过度表达,从而导致慢性炎症和组织损伤。因此,在本研究中,硒代蛋氨酸抑制了大肠杆菌诱导的 TNF-α 和 IL-1β 蛋白表达,表明硒代蛋氨酸抑制大肠杆菌感染引起的奶牛乳腺上皮细胞和巨噬细胞炎症,从而抑制了炎性风暴。

炎症信号通路的激活已被证实在调节免疫应答中发挥重要作用。大肠杆菌感染导致乳房 TLR4/NF-κB 信号通路的激活。因此,大肠杆菌感染诱导的 TLR4、IκB、TNF-α、IL-1β 蛋白表达和 NF-κB 核蛋白表达显著升高,这表明大肠杆菌激活了奶牛乳腺上皮细胞和巨噬细胞中 TLR4/NF-κB 信号通路,这与大肠杆菌通过 TLR4/NF-κB 途径引起乳腺炎的研究结果一致。不同浓度的硒代蛋氨酸增加了硒蛋白 S 的表达,而降低了奶牛乳腺上皮细胞和巨噬细胞中 TLR4、IκB、TNF-α、IL-1β 蛋白表达和 NF-κB 核蛋白表达,这与硒补充剂在小鼠乳腺上皮细胞中通过 NF-κB 信号通路抑制脂多糖诱导的炎症因子增加的研究结果保持一致。此外,硒缺乏通过调节小鼠乳腺和巨噬细胞内的 TLR2/NF-κB 通路加重了金黄色葡萄球菌感染后的炎症。因此,硒代蛋氨酸在奶牛乳腺上皮细胞和巨噬细胞中通过促进硒蛋白 S 的表达抑制 TLR4 介导的 NF-κB 信号通路缓解大肠杆菌感染诱导的炎性反应。

硒蛋白是硒蛋白家族的一员,在多种组织中表达,参与细胞应激反应、免疫和炎症过程。在本研究中,硒蛋白 S 敲除部分逆转了大肠杆菌感染诱导的 TLR4、IKB、NF-κB、TNF-α 和 IL-1β 蛋白表达的升高。因此,硒蛋白 S 敲除通过介导 TLR4/NF-κB 信号通路抑制 TNF-α 和 IL-1β 的表达。有趣的是,本研究明确了硒代蛋氨酸在硒蛋白 S 敲除的奶牛乳腺上皮细胞和巨噬细胞中不能完全抑制 TLR4、IKB、NF-κB、TNF-α 和 IL-1β 的表达。因此,硒蛋白 S 在硒代蛋氨酸通过 TLR4/NF-κB 通路缓解大肠杆菌感染诱导的奶牛乳腺上皮细胞和巨噬细胞炎性反应中具有重要作用。此外,在大肠杆菌和/或硒代蛋氨酸存在的情况下,NF-κB 抑制剂 BAY11-708 不影响硒蛋白 S 的表达,但抑制了 TNF-α 和 IL-1β 的表达,这表明硒蛋白 S 和抑制 NF-κB 均能减轻大肠杆菌感染诱导的炎性反应。此外,虽然硒蛋白 S 是 NF-κB 的靶基因,但额外补充硒可以促进更多的硒蛋白 S 表达,硒蛋白 S 水平的增加对 NF-κB 的激活具有负反馈作用。因此,硒代蛋氨酸通过促进硒蛋白 S 表达,抑制 TLR4/NF-κB 信号通路,进而抑制大肠杆菌感染诱导的炎性反应。

有趣的是,大肠杆菌感染在 4 h 内显著促进了奶牛乳腺上皮细胞中 TLR4 介导的 NF-κB 信号通路相关蛋白的表达,而在巨噬细胞中这些蛋白发生变化需 6 h,这个差异可能是由于大肠杆菌先侵入奶牛乳腺上皮细胞,然后促进趋化因子的表达,这些趋化因子会招募并激活巨噬细胞。最后,巨噬细胞吞噬大肠杆菌等致病菌,并引发炎性反应。

九、结论

大肠杆菌诱导奶牛乳腺上皮细胞中趋化因子的表达,然后招募和激活巨噬细胞。此外,硒代蛋氨酸在奶牛乳腺上皮细胞和巨噬细胞中通过激活硒蛋白 S 介导的 TLR4/NF-κB 信号通路,减轻了大肠杆菌诱导的炎症(图 3-15),为硒的细胞保护作用提供了实验依据。

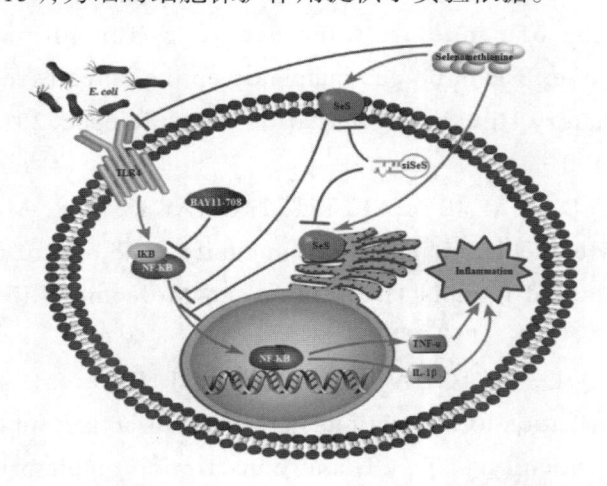

图 3-15 硒代蛋氨酸对大肠杆菌感染引起的炎症的保护作用

参考文献

[1] OLDE RIEKERINK R G M, BARKEMA H W, KELTON D F, et al. Incidence rate of clinical mastitis on Canadian dairy farms[J]. Journal of Dairy Science, 2008, 9: 1366-1377.

[2] GAO J, BARKEMA H W, ZHANG L M, et al. Incidence of clinical mastitis and distribution of pathogens on large Chinese dairy farms[J]. Journal of Dairy Science, 2017, 100（6）: 4797-4806

[3] VANGROENWEGHE F, LAMOTE I, BURVENICH C. Physiology of the periparturient period and its relation to severity of clinical mastitis[J]. Domestic Animal Endocrinology, 2005, 29（2）: 283-293.

[4] GILBERT F B, CUNHA P, JENSEN K, et al. Differential response of bovine mammary epithelial cells to Staphylococcus aureus or Escherichia coli agonist s of the innate immune system[J]. Veterinary Research, 2013, 44（1）: 40.

[5] BOUGARN S, CUNHA P, GILBERT F B, et al. Technical note: Validation of candidate reference genes for normalization of quantitative PCR in bovine mammary epithelial cells responding to inflammatory stimuli[J]. Journal of Dairy Science, 2011, 94（5）: 2425-2430.

[6] SWAMYDAS M, BREAK T J, LIONAKIS M S. Mononuclear phagocyte-mediated antifungal immunity: The role of chemotactic receptors and ligands[J]. Cellular and Molecular Life Sciences, 2015, 72（11）: 2157-2175.

[7] LECOQ L, RAIOLA L, CHABOT P R, et al. Structural characterization of interactions between transactivation domain 1 of the p65 subunit of NF-κB and transcription regulatory factors[J]. Nucleic Acids Research, 2017, 45（9）: 5564-5576.

[8] SALMAN S, KHOL-PARISINI A, SCHAFFT H, et al. The role of dietary selenium in bovine mammary gland health and immune function[J]. Animal Health Research Reviews, 2009, 10（1）: 21-34.

[9] KUBACHKA K M, HANLEY T, MANTHA M, et al. Evaluation of selenium in dietary supplements using elemental speciation[J]. Food Chemistry, 2017, 218: 313-320.

[10] KAUSHAL N, KUDVA A K, PATTERSON A D, et al. Crucial role of macrophage selenoproteins in experimental colitis[J]. Journal of Immunology, 2014, 193 (7): 3683-3692.

[11] SHIMOHASHI N, NAKAMUTA M, UCHIMURA K, et al. Selenoorganic compound, ebselen, inhibits nitric oxide and tumor necrosis factor-α production by the modulation of jun-N-terminal kinase and the NF-κB signaling pathway in rat Kupffer cells[J]. Journal of Cellular, 2000, 78 (4): 595-606.

[12] CEBRA C K, HEIDEL J R, CRISMAN R O, et al. The relationship between endogenous cortisol, blood micronutrients, and neutrophil function in postparturient Holstein cows[J]. Journal of Veterinary Internal Medicine, 2003, 17 (6): 902-907.

[13] KIEŁCZYKOWSKA M, KOCOT J, PAZDZIOR M, et al. Selenium-a fascinating antioxidant of protective properties[J]. Advances in Clinical and Experimental Medicine, 2018, 27 (2): 245-255.

[14] FENG R, DESBORDES S C, XIE H F, et al. PU 1 and C/EBPalpha/ beta convert fibroblasts into macrophage-like cells[J]. Proceedings of the National Academy of Sciences of the United States of America, 2008, 105 (16): 6057-6062.

[15] ZOU Y X, SHAO J J, LI Y X, et al. Protective effects of inorganic and organic selenium on heat stress in bovine mammary epithelial cells[J]. Oxidative Medicine and Cellular Longevity, 2019, 2019: 1503478.

[16] WEISS W P, HOGAN J S. Effect of selenium source on selenium status, neutrophil function, and response to intramammary endotoxin challenge of dairy cows[J]. Journal of Dairy Science, 2005, 88 (12): 4366-4374.

[17] KIM Y Y, MAHAN D C. Comparative effects of high dietary levels of organic and inorganic selenium on selenium toxicity of

growing-finishing pigs[J]. Journal of Animal Science, 2001, 79 (4): 942-948.

[18] JIN Y, YE X, SHAO L, et al. Serum lactic dehydrogenase strongly predicts survival in metastatic na sopharyngeal carcinoma treated with palliative chemotherapy[J]. European Journal of Cancer, 2013, 49 (7): 1619-1626.

[19] ARIBI M, MEZIANE W, HABI S, et al. Macrophage bactericidal activities against staphylococcus aureus are enhanced in vivo by selenium supplementation in a dose-dependent manner[J]. PLoS One, 2015, 10 (9): e0135515.

[20] PORCHERIE A, CUNHA P, TROREREAU A, et al. Repertoire of Escherichia coli agonists sensed by innate immunity receptors of the bovine udder and mammary epithelial cells[J]. Veterinary Research, 2012, 43 (1): 14-21.

[21] SCHEIBEL M, KLEIN B, MERKLE H, et al. IκBβ is an essential coactivator for LPS-induced IL-1β transcription in vivo[J]. Journal of Experimental Medicine, 2010, 207 (12): 2621-2630.

[22] YANG W, ZERBE H, PETZL W, et al. Bovine TLR2 and TLR4 properly transduce signals from Staphylococcus aureus and E. coli, but S. aureus fails to both activate NF-kappaB in mammary epithelial cells and to quickly induce TNFalpha and interleukin-8 (CXCL8) expression in the udder[J]. Molecular Immunology, 2008, 45 (5): 1385-1397.

[23] MA N N, CHANG G J, HUANG J, et al. Cis-9, trans-11-Conjugated Linoleic Acid Exerts an Anti-inflammatory Effect in Bovine Mammary Epithelial Cells after Escherichia coli Stimulation through NF-κB Signaling Pathway[J]. Journal of Agricultural and Food Chemistry, 2019, 67 (1): 193-200.

[24] ZHANG W, ZHANG R X, WANG T C, et al. Selenium inhibits LPS-induced pro-inflammatory gene expression by modulating MAPK and NF-κB signaling pathways in mouse mammary epithelial cells in primary culture[J]. Inflammation, 2014, 37 (2):

478-485.

[25] GAO X J, ZHANG Z C, LI Y, et al. Selenium deficiency facilitates inflammation following S aureus infection by regulating TLR2-related pathways in the mouse mammary gland[J]. Biological Trace Element Research, 2016, 172 (2): 449-457.

[26] XU J W, GONG Y F, SUN Y, et al. Impact of selenium deficiency on inflammation, oxidative stress, and phagocytosis in mouse macrophages[J]. Biological Trace Element Research, 2020, 194 (1): 237-243.

[27] SANTOS L R, DURÃES C, ZIROS P G, et al. Interaction of genetic variations in NFE2L2 and SELENOS modulates the risk of hashimoto's thyroiditis[J]. Thyroid, 2019, 29 (9): 1302-1315.

[28] GAO Y, HANNAN N R F, WANYONYI S, et al. Activation of the selenoprotein SEPS1 gene expression by pro-inflammatory cytokines in HepG2 cells[J]. Cytokine, 2006, 33 (5): 246-251.

第三节 硒缓解大肠杆菌感染致细胞凋亡的机制研究

一、概述

哺乳期乳腺经常受到传染性和环境病原体的侵袭,可诱发牛乳腺炎。大肠杆菌作为一种环境致病菌,可引起奶牛泌乳早期的临床乳腺炎。细胞凋亡在乳腺炎的发生发展中起着重要作用。虽然细胞凋亡是维持机体组织内稳态的一种生理机制,但一些病原体通过诱导细胞凋亡导致组织破坏进而对宿主免疫应答产生不利影响。细胞凋亡的信号传导途径分为外源性(死亡受体)和内源性(线粒体)途径。在外源性途径中,定位于各种细胞表面的死亡受体 Fas 与其配体 FasL 相互作用,激活 caspase 8。此外,在内源性途径中,线粒体对凋亡信号的反应受到 Bcl-2 家族的高度调控,这个家族也包括抗凋亡成员(Bcl-2、Bcl-xl 和 Bcl-W 等)和促凋亡成员(Bax、Bak 和 Bim)。它们可以通过调节线粒体通透性过渡孔来控制促凋亡蛋白(细胞色素 C,CytoC)从线粒体释放到细胞质中,然后,CytoC 通过 APAF1 募集并激活 caspase 9,最后,caspase 8 和/或 caspase 9 激活 caspase 3,诱导细胞凋亡。

硒是一种必需的矿物质营养素。有机硒,特别是硒代蛋氨酸是哺乳动物硒的主要膳食来源。动物缺乏硒是一个全球性的问题,它增加了对各种疾病的易感性,降低了生产力。已有研究表明,亚硒酸钠对脂多糖诱导的 MC3T3-E1 细胞凋亡有保护作用,且硒对糖尿病诱导的大鼠背根神经节和海马细胞凋亡有抑制作用。然而,硒对大肠杆菌感染奶牛乳腺上皮细胞的保护作用和凋亡机制尚不清楚。因此,我们利用大肠杆菌感染奶牛乳腺上皮细胞的体外实验模型,研究硒是否通过调节死亡受体和/或线粒体途径发挥抗凋亡作用。

二、硒代蛋氨酸抑制大肠杆菌诱导的奶牛乳腺上皮细胞细胞凋亡

通过 FAC 流式细胞术检测细胞凋亡（图 3-16）。与对照组相比，大肠杆菌感染组奶牛乳腺上皮细胞细胞凋亡率显著升高（$P < 0.01$），而硒代蛋氨酸预处理可显著缓解这种升高（$P < 0.01$）。在大肠杆菌感染的奶牛乳腺上皮细胞中，硒蛋白 S 敲除可抵消硒代蛋氨酸降低大肠杆菌感染诱导细胞凋亡增加的作用（$P < 0.01$），而 caspase 8 抑制剂 ZIK 和细胞凋亡抑制剂 Nec-2S 均可显著抑制大肠杆菌诱导的细胞凋亡（$P < 0.05$）。

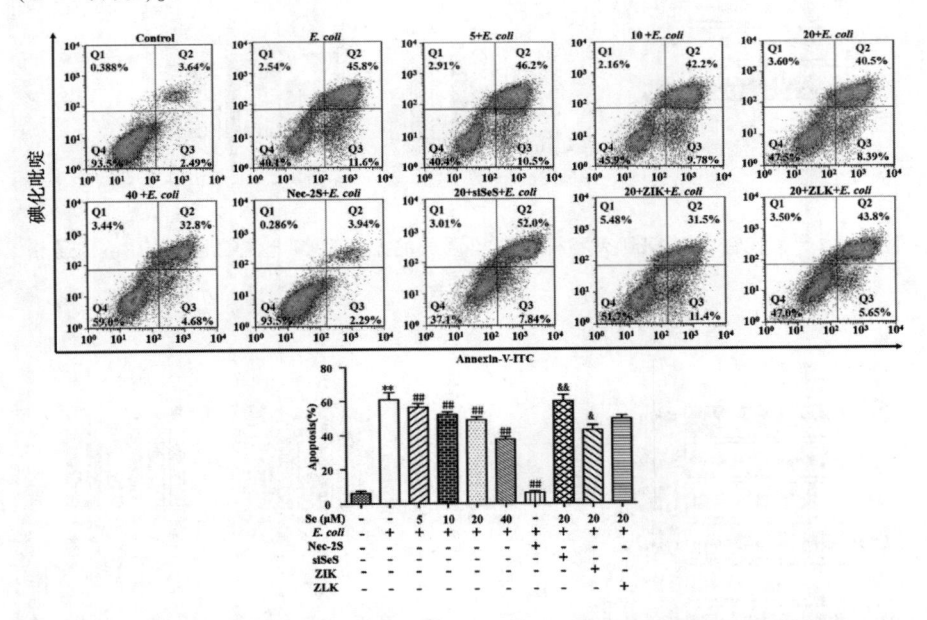

图 3-16 硒代蛋氨酸对大肠杆菌感染诱导的奶牛乳腺上皮细胞凋亡的保护作用

三、硒代蛋氨酸增加了硒蛋白 S 的表达且抑制了 Fas/FasL 和线粒体凋亡通路的激活

硒代蛋氨酸缓解了大肠杆菌感染诱导的奶牛乳腺上皮细胞中硒蛋白 S 和细胞膜电位的下调（$P < 0.05$，图 3-17A，B 和图 3-19）。对于 Fas/FasL 和线粒体途径，大肠杆菌感染显著增加了 Fas、FasL、FADD、cleaved-caspase、细胞质细胞色素 C（C-CytoC）、cleaved-caspase 9

和 cleaved-caspase 3 的蛋白表达（$P < 0.05$），且显著降低了 Bcl-2 与 Bax 蛋白表达的比值、线粒体细胞色素 C（M-CytoC）蛋白表达及细胞膜电位（$P < 0.01$）；而硒代蛋氨酸预处理可拮抗这些变化（图 3-17 ~ 图 3-19）。

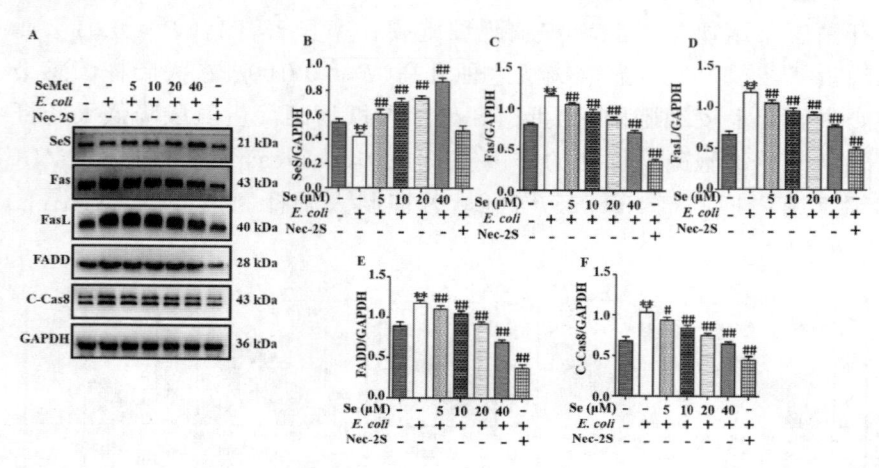

图 3-17 硒代蛋氨酸抑制大肠杆菌感染诱导的奶牛乳腺上皮细胞 Fas/FasL 通路的激活

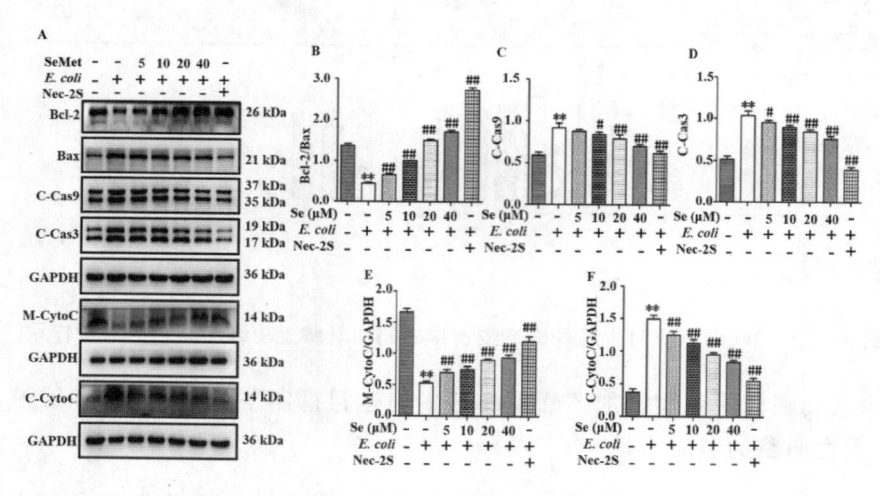

图 3-18 硒代蛋氨酸抑制大肠杆菌感染诱导的奶牛乳腺上皮细胞线粒体通路的激活

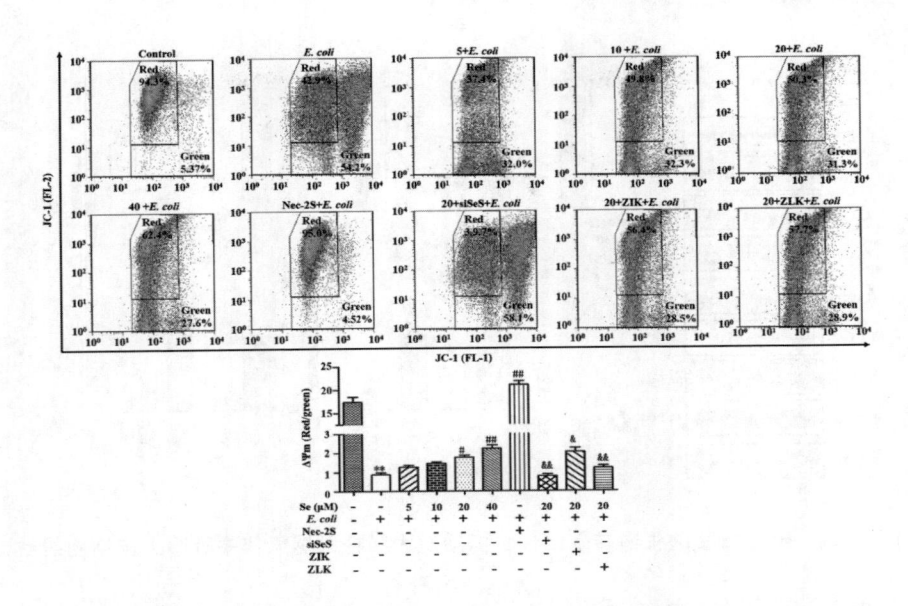

图 3-19　硒代蛋氨酸抵消了大肠杆菌感染奶牛乳腺上皮细胞线粒体膜电位的下降

四、硒蛋白 S 缺失对 Fas/FasL 和线粒体通路激活的缓解作用

在大肠杆菌感染和硒代蛋氨酸共同处理的奶牛乳腺上皮细胞中，敲低硒蛋白 S 可显著缓解线粒体膜电位，以及 Fas、FasL、FADD、cleaved-caspase 8、C-CytoC、cleaved-caspase 9 和 cleaved-caspase 3 蛋白表达的降低（$P<0.01$），同时降低 Bcl-2 和 Bax 比值以及 M-CytoC 蛋白表达的升高（$P<0.01$，图 3-20、图 3-21）。

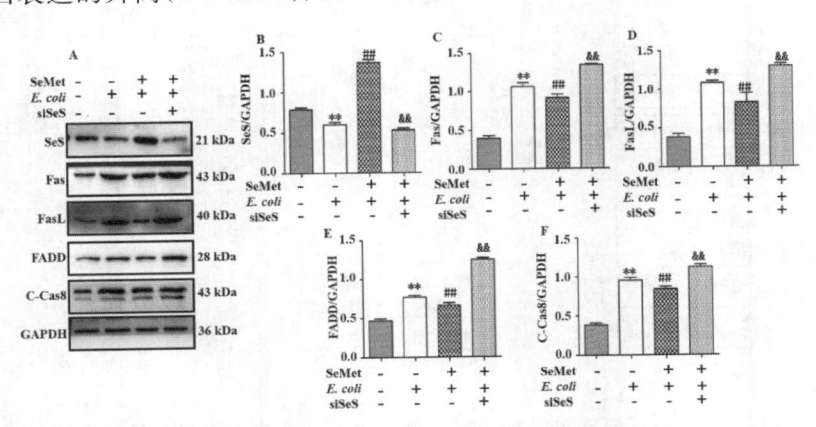

图 3-20　硒蛋白 S 缺失逆转了硒代蛋氨酸对大肠杆菌感染奶牛乳腺上皮细胞 Fas/FasL 通路的影响

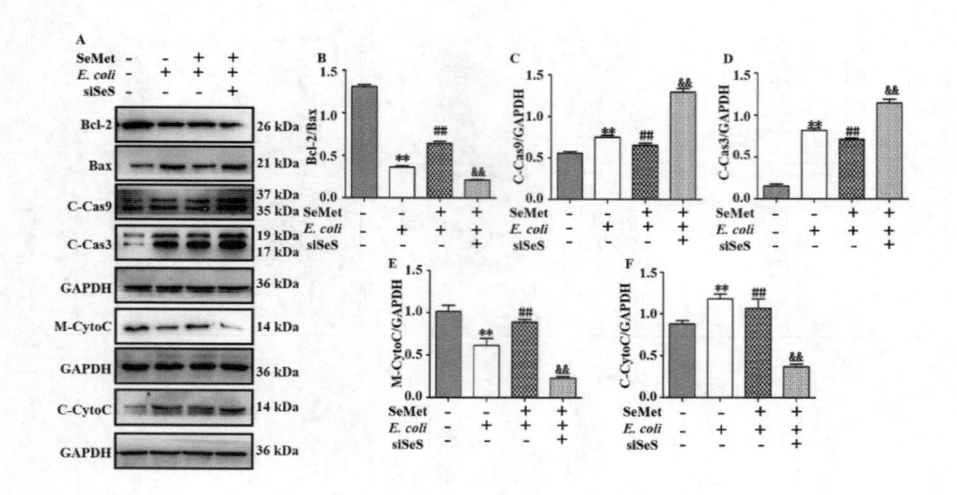

图 3-21 硒蛋白 S 缺失逆转了硒代蛋氨酸对大肠杆菌感染奶牛乳腺上皮细胞线粒体通路的影响

五、Fas/FasL 通路在硒代蛋氨酸缓解大肠杆菌感染奶牛乳腺上皮细胞中的作用

在大肠杆菌感染和硒代蛋氨酸（20 μM）处理的奶牛乳腺上皮细胞中，caspase 8 抑制剂 ZIK 显著抑制了 cleaved-caspase 8、C-CytoC、cleaved-caspase 9 和 cleaved-caspase 3 蛋白水平（$P < 0.05$），且显著增加了 Bcl-2/Bax 比值（$P < 0.05$，图 3-22），显著抑制了细胞凋亡（$P < 0.01$，图 3-17），增加了线粒体膜电位（$P < 0.01$，图 3-19）。

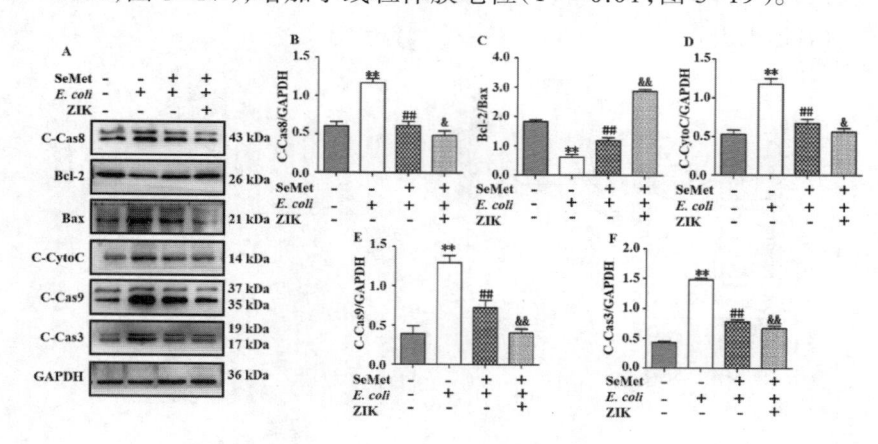

图 3-22 Fas/FasL 通路参与硒代蛋氨酸对大肠杆菌感染诱导的奶牛乳腺上皮细胞凋亡的调控

六、线粒体途径对细胞凋亡过程的影响

caspase 9 抑制剂 ZLK 显著抑制大肠杆菌感染和硒代蛋氨酸处理组 cleaved-caspase 9、Fas、FasL 和 FADD 蛋白水平（$P<0.05$，图 3-23）。

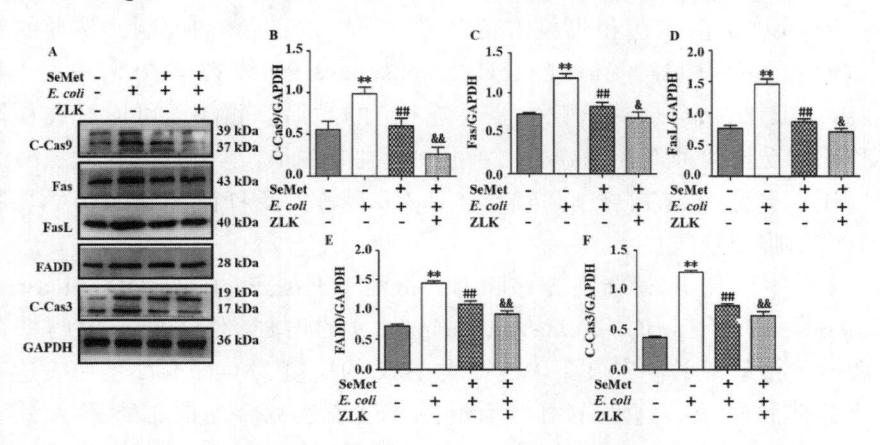

图 3-23 线粒体途径参与了硒代蛋氨酸对大肠杆菌感染诱导的奶牛乳腺上皮细胞凋亡的调控

七、讨论

5 μM、10 μM、20 μM 和 40 μM 硒代蛋氨酸预处理，显著改善了大肠杆菌感染奶牛乳腺上皮细胞中硒蛋白 S 蛋白表达的减少。硒是一种含缬草苷的蛋白相互作用膜蛋白，它的状态会改变。高硒饲粮摄入量增加了猪肝脏和肌肉中硒蛋白 S 的基因表达。在本研究中，大肠杆菌感染降低了硒蛋白 S 的蛋白表达，而所有浓度的硒代蛋氨酸都促进了硒蛋白 S 的表达，并抵消了大肠杆菌诱导的奶牛乳腺上皮细胞中硒蛋白 S 的降低。

在本研究中，不同浓度的硒代蛋氨酸和凋亡抑制剂 Nec-2S，显著降低了大肠杆菌感染的奶牛乳腺上皮细胞中 cleaved-caspase3 蛋白的表达和凋亡率，这与硒代蛋氨酸降低氧化应激下奶牛乳腺上皮细胞凋亡保持一致。硒能降低 2,4-二硝基氯苯诱导的奶牛乳腺上皮细胞细胞凋亡活化。因此，在本研究中，硒代蛋氨酸抑制了大肠杆菌感染的奶牛乳

腺上皮细胞的凋亡。

细胞凋亡主要通过两种途径诱导,死亡受体的激活(外源性途径)或线粒体的激活(内在途径)。在本研究中,硒代蛋氨酸处理显著抑制了奶牛乳腺上皮细胞中 Fas、FasL、FADD、cleaved-caspase 8、C-CytoC、cleaved-caspase 9 的蛋白表达,而显著增加了 Bcl-2 与 Bax 的比值、M-CytoC 蛋白表达和线粒体膜电位。同样,亚硒酸钠抑制铅诱导的鸡中性粒细胞 Bcl-2/Bax-CytoC-caspasepase 9(线粒体凋亡途径)和 Fas-FADD-caspase 8(死亡受体途径)的激活。硒缺乏可通过线粒体和死亡受体两种途径诱导十二指肠绒毛细胞凋亡。因此,硒代蛋氨酸通过死亡受体(Fas/FasL)和线粒体途径,缓解大肠杆菌诱导的奶牛乳腺上皮细胞细胞凋亡。

在本研究中,硒蛋白 S 敲除部分抵消了 Fas、FasL、FADD、cleaved-caspase8、C-CytoC 和 cleaved-caspase 9 的增加,且降低了大肠杆菌和硒代蛋氨酸共同存在下 Bcl-2 与 Bax 的比值、M-CytoC 蛋白表达和线粒体膜电位。这些作用与硒蛋白 S 过表达减轻赭曲霉毒素 A 诱导 PK15 细胞凋亡一致,而硒蛋白 S 敲低则加重了赭曲霉毒素 A 诱导的 PK15 细胞凋亡。因此,硒蛋白 S 敲除通过介导 Fas/FasL 和线粒体途径激活奶牛乳腺上皮细胞凋亡。有趣的是,本研究明确了硒代蛋氨酸在硒蛋白 S 缺失的奶牛乳腺上皮细胞中不抑制 Fas、FasL、FADD、cleaved-caspase8、CytoC 和 cleaved-caspase9 的表达,但它促进了 Bcl-2 与 Bax 的比值。因此,硒蛋白 S 在硒代蛋氨酸通过死亡受体和线粒体通路,缓解大肠杆菌感染诱导的奶牛乳腺上皮细胞凋亡中具有重要作用。

ZIK 是 caspase 8 抑制剂,而 ZLK 显著抑制了 caspase 9。在本研究中,大肠杆菌、硒代蛋氨酸和 ZIK 处理的奶牛乳腺上皮细胞中,cleaved-caspase8、CytoC、cleaved-caspase9 和 cleaved-caspase3 的蛋白表达和凋亡率显著低于仅大肠杆菌和硒代蛋氨酸处理的细胞,Bcl-2 与 Bax 的比值显著高于仅大肠杆菌和硒代蛋氨酸处理的细胞。此外,ZLK 在大肠杆菌和硒代蛋氨酸存在下显著抑制 Fas、FasL、FADD 和 Cleaved-caspase9 蛋白的表达。Caspase 9 是线粒体凋亡途径中的启动物,而 caspase-8 在死亡受体介导的凋亡激活的调控和启动中是不可或缺的。因此,我们推断硒代蛋氨酸和 ZIK/ZLK 具有协同作用,可以减少大肠杆菌诱导的细胞凋亡。

八、结论

硒代蛋氨酸通过激活硒蛋白S介导的线粒体和Fas/FasL(死亡受体)通路缓解大肠杆菌感染诱导奶牛乳腺上皮细胞的凋亡(图3-24)。因此,硒对大肠杆菌感染的奶牛乳腺上皮细胞具有细胞保护作用。

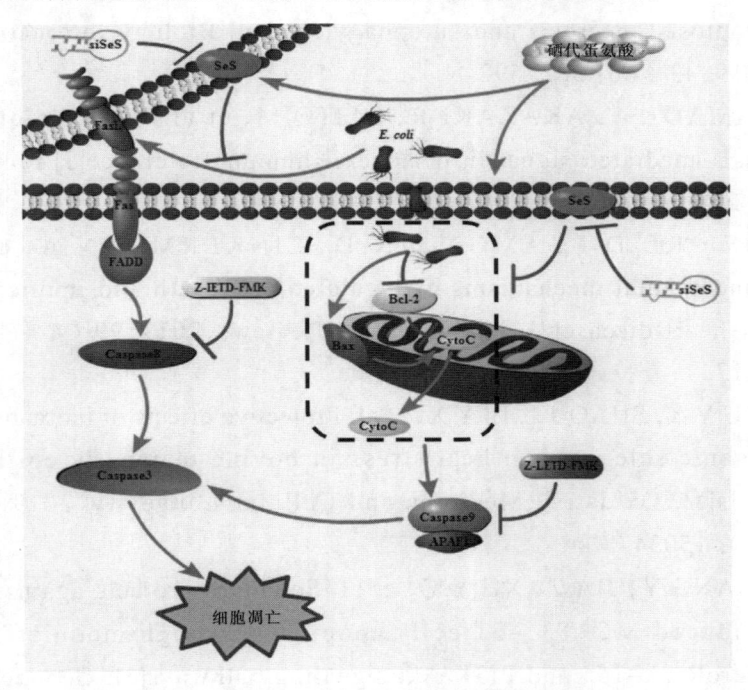

图3-24 硒代蛋氨酸缓解大肠杆菌感染诱导的奶牛乳腺上皮细胞凋亡

参考文献

[1] CALDEIRA M O, BRUCKMAIER R M, WELLNITZ O. Meloxicam affects the inflammatory responses of bovine mammary epithelial cells[J].Journal of Dairy Science, 2019, 102(11): 10277-10290.

[2] GAO J, BARKEMA H W, ZHANG L M, et al. Incidence of clinical mastitis and distribution of pathogens on large Chinese dairy farms[J]. Journal of Dairy Science, 2017, 100(6): 4797-4806.

[3] SHAHID M, GAO J, ZHOU Y N, et al. Prototheca zopfii isolated

from bovine mastitis induced oxidative stress and apoptosis in bovine mammary epithelial cells[J]. Oncotarget, 2017, 8 (19): 31938-31947.

[4] DELEO F R. Modulation of phagocyte apoptosis by bacterial pathogens[J]. Apoptosis, 2004, 9 (4): 399-413.

[5] D'ARCY M S. Cell death: A review of the major forms of apoptosis, necrosis and autophagy[J]. Cell Biology International, 2019, 43 (6): 582-592.

[6] YAMADA A, AKAKAKI R, SAITO M, et al. Dual role of Fas/FasL-mediated signal in peripheral immune tolerance[J]. Frontiers in Immunology, 2017, 8: 403.

[7] DEL RE D P, AMGALAN D, LINKERMANN A, et al. Fundamental mechanisms of regulated cell death and implications for heart disease[J]. Physiological Reviews, 2019, 99 (4): 1765-1817.

[8] ZOU Y X, SHAO J J, LI Y X, et al. Protective effects of inorganic and organic selenium on heat stress in bovine mammary epithelial cells[J]. Oxidative Medicine and Cellular Longevity, 2019, 2019 (1): 1503478.

[9] HUANG Y, JIA Z, XU Y Q, et al. Selenium protects against LPS-induced MC3T3-E1 cells apoptosis through modulation of microRNA-155 and PI3K/Akt signaling pathways[J]. Genetics and Molecular Biology, 2020, 43 (3): e20190153.

[10] KAHYA M C, NAZIROĞLU M, ÖVEY İ S. Modulation of diabetes-induced oxidative stress, apoptosis, and Ca (2+) entry through TRPM2 and TRPV1 channels in dorsal root ganglion and hippocampus of diabetic rats by melatonin and selenium[J]. Molecular Neurobiology, 2017, 54 (3): 2345-2360.

[11] COMINETTI C, DE BORTOLI M C, PURGATTO E P, et al. Associations between glutathione peroxidase-1 Pro198Leu polymorphism, selenium status, and DNA damage levels in obese women after consumption of Brazil nuts[J]. Nutrition, 2011, 27 (9): 891-896.

[12] ZHAO Z, BARCUS M, KIM J, et al. High dietary selenium intake alters lipid metabolism and protein synthesis in liver and muscle of pigs[J]. The Journal of Nutrition, 2016, 146(9): 1625-1633.

[13] MIRANDA S G, PURDIE N G, OSBORNE V R, et al. Selenomethionine increases proliferation and reduces apoptosis in bovine mammary epithelial cells under oxidative stress[J]. Journal of Dairy Science, 2011, 94(1): 165-173.

[14] GUO Y M, YAN S M, GONG J, et al. The protective effect of selenium on bovine mammary epithelial cell injury caused by depression of thioredoxin reductase[J]. Biological Trace Element Research, 2018, 184(1): 75-82.

[15] ELMORE S. Apoptosis: A review of programmed cell death[J]. Toxicologic Pathology, 2007, 35(4): 495-516.

[16] YIN K, CUI Y, SUN T, et al. Antagonistic effect of selenium on lead-induced neutrophil apoptosis in chickens via miR-16-5p targeting of PiK3R1 and IGF1R[J]. Chemosphere, 2020, 246: 125794.

[17] WANG J F, LIU Z, HE X J, et al. Selenium deficiency induces duodenal villi cell apoptosis via an oxidative stress-induced mitochondrial apoptosis pathway and an inflammatory signaling-induced death receptor pathway[J]. Metallomics, 2018, 10(10): 1390-1400.

[18] GAN F, HU Z H, ZHOU Y J, et al. Overexpression and low expression of selenoprotein S impact ochratoxin A-induced porcine cytotoxicity and apoptosis in vitro[J]. Journal of Agricultural and Food Chemistry, 2017, 65(32): 6972-6981.

[19] BRENTN ALL M, RODRIGUEZ-MENOCAL L, DE GUEVARA R L, et al. Caspase-9, caspase-3 and caspase-7 have distinct roles during intrinsic apoptosis[J]. BMC Cell Biology, 2013, 14: 32.

[20] TUMMERS B, GREEN D R. Caspase-8: regulating life and death[J]. Immunological Reviews, 2017, 277(1): 76-89.

后 记

奶业是我国畜牧业发展的支柱性产业。奶牛乳腺炎作为奶牛常见的疾病之一,严重地影响奶业的发展,造成了巨大的经济损失。据报道,我国每年因奶牛乳腺炎导致的经济损失高达 30 亿元人民币。大肠杆菌作为奶牛乳腺炎重要的致病菌之一,不但会引起多个系统疾病(如胃肠道、尿道、关节等),还特别容易引发奶牛乳腺炎。目前,抗生素是国内外治疗奶牛乳腺炎最常用的方法,然而其大量不合理的使用甚至滥用随之带来了耐药致病菌株的出现和传播、畜产品抗生素残留等问题,威胁着人类公共卫生安全。β–内酰胺类药物是兽医治疗大肠杆菌型奶牛乳腺炎的首选药物,这使产超广谱 β–内酰胺酶大肠杆菌的发生率逐渐增加,无疑增加了大肠杆菌型奶牛乳腺炎的治疗难度。因此,寻找抗生素的替代物已经迫在眉睫。

依据国家统计局所公布的数据,在 2022 年,全国肉类产量达 8990 万吨,奶类产量为 3778 万吨,禽蛋产量是 3409 万吨,多年来畜牧业的持续发展有力保障了肉蛋奶的稳定供应,期待本书的出版为大肠杆菌型奶牛乳腺炎的防治提供依据。

本书及其研究内容是由我、我的老师、我的同学和我的学生们共同完成的。自 2017 年开始,参与的老师有韩博教授、高健副教授、王东研究员、John P. Kastelic 教授、Herman W. Barkema 教授和 Sadeeq Ur Rahman 教授,参与的同学们有张诗瑶、Muhammad Shahid、周曼、刘钢、杨晶越、霍文琳、徐司雨和刘洋,参与的学生们有赵晋辉、张新颖、白健刚、张若晴和杨天宇。

本书是我的阶段性研究成果,一些结论是基于本课题组的研究得出的,是一家之言,不妥之处,期望各位同行、专家和读者多多赐教。学海无涯、永无止境,由于本人学识有限,该领域的很多问题有待于进一步研究和深化,我将继续努力探索和学习。

本书的出版要感谢方方面面的支持。感谢山西省教育厅、山

后 记

西省科技厅、山西省基础研究计划自由探索类青年科学研究项目：FXR-FGF15介导的胆汁酸循环在氟致肝脂代谢紊乱中的作用机制（202203021212434）和山西农业大学博士科研启动项目：呕吐毒素影响肉鸡肌肉发育的机制（2021BQ73）对研究的支持和资助。感谢山西农业大学提供的优越科研环境，以及动物医学院、临床兽医学科先进的分子生物学实验室及其实施的共享制度。特别感谢我的老师和同学们在科研方面的辛勤付出，以上都是我科研的坚强后盾，使本研究得以持续顺利进行。在本书形成过程中，本人参考了国内外同行专家和学者的部分研究成果，在此表示衷心的感谢！

庄翠翠
2024年7月于山西农业大学